The Distribution of Australian Dragonflies

Ian Endersby

Layout and typesetting: Busybird Publishing

Busybird Publishing
2/118 Para Road
Montmorency, Victoria
Australia 3094
www.busybird.com.au

Cover: Distribution of *Orthetrum caledonicum*

Frontispiece:

(Upper)   Most speciose IBRA7 region [145 species] (Wet Tropics)

(Lower)   IBRA7 region with most records [8,957] (Southeast Queensland)

Dedicated to Margaret Rose Endersby (1939 – 2014)
whose favourite species was *Orthetrum caledonicum*

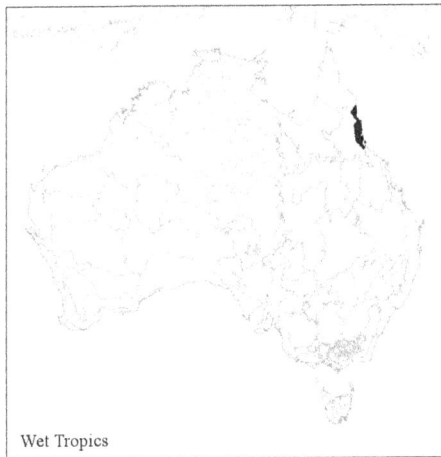
Wet Tropics

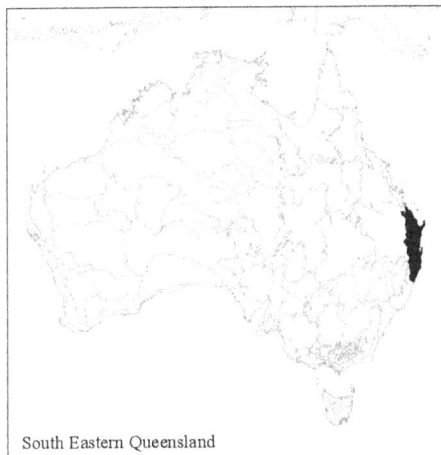
South Eastern Queensland

Contents

Preface

Upon reading in Tony Watson's paper[1] on the distribution of Australia's dragonflies that Victoria (my home state) had only 63 species, I realised that it was a number I could get my mind around. Not like the beetles, moths or flies. Soon after, the chance finding of Fraser's handbook[2] of Australasian dragonflies in a secondhand book shop enabled me to extract a key to the Victorian species; thus was a new interest sparked.

Having seen many of the local species, observed prolonged tandem underwater oviposition and the temperature-induced colour change of *Austrolestes annulosus*, I wondered if it would be possible to map the distribution of the Victorian species. So I visited the collection of the Melbourne Museum only to find that the label data had not yet been digitised. Years later, after re-gluing many dislodged heads and legs, I completed that task for the whole collection, which was Australia wide. With that as a start the project burgeoned to become mapping the distribution of all species recorded within Australia. With the invaluable help of curators and collection managers I was able to assemble all of the digitised records from Australian museums. Hobart required a special visit to digitise its collection label data. A preliminary set of maps was published in updated identification key[3] prepared by Gunther Theischinger and myself.

Since then I have continued to gather distribution data from overseas museums, additional records of Australian museums, species lists of visitors who can authenticate their identifications, from amongst the plethora of wildlife photographers those whose identifications I

---

1. Watson, J.A.L. (1974) The distributions of the Australian Dragonflies (Odonata). *Journal of the Australian Entomological Society* 13: 137-149

2. Fraser, F.C. (1960) *A handbook of the dragonflies of Australasia, with keys for Identification of all species.* Royal Zoological Society of New South Wales: Sydney. 67 pp. 26 pls.

3. Theischinger, G. & Endersby, I. (2009) *Identification Guide to the Australian Odonata.* Department of Environment, Climate Change and Water, NSW

can trust, and where there are big gaps, from the odonatalogical literature.

Encouraged by Vincent Kalkman to use the data to produce charts of flight times, it was obvious from the size and geographic spread of Australia, that any such charts would be latitude-, if not altitude-, dependent. The Interim Biogeographic Regionalisation for Australia (currently IBRA7) classifies Australia's landscapes into 89 large geographically distinct bioregions based on common climate, geology, landform, native vegetation and species information. It seemed an ideal vehicle to overcome this problem and so the distribution of each species was also sorted into IBRA7 categories before flight times were analysed. However, the characteristics which define biogeographic regions are probably not those which influence dragonfly distribution so a further analysis using climate zones was also made.

And so you have this book.

Sources

Australian Museum, Sydney, New South Wales

Australian National Insect Collection, Division of Entomology, CSIRO, Canberra City, A.C.T.

California State Collection of Arthropods. California Department of Food and Agriculture. USA

Dennis Paulson private collection

Environment Protection Authority (EPA) NSW, Sydney, New South Wales

Fons Peel Dragonflypix photographic collection

Gunther Theischinger private collection

Museum and Art Galleries of the Northern Territory, Darwin, Northern Territory

Museum of Victoria, Melbourne, Victoria

Nationaal Natuurhistorisch Museum (formerly Rijksmuseum van Natuurlijke Historie), Leiden, Netherlands - Vincent Kalkman collections

National Museum of Natural History (Smithsonian Institution), Washington, D.C., USA

Office of *Environment* and Heritage New South Wales - Data from Scientific Licences (*Petalura gigantea*)

Queen Victoria Museum, Launceston, Tasmania

Queensland Museum, Brisbane, Queensland

Reiner Richter from Atlas of Living Australia 2018-05-04

Rosser W. Garrison private collection

South Australian Museum, Adelaide, South Australia

Tasmanian Museum and Art Gallery, Hobart, Tasmania

Western Australian Museum, Perth, Western Australia

Scientific Literature

Acknowledgements

For provision of specimen distribution data: Tom Weir, Federica Turco, Jaime Florez (ANIC); David Britton, Derek Smith (Australian Museum); Ken Walker, Peter Lillywhite (Melbourne Museum); Jan Forrest (South Australian Museum); Gavin Dally (Museums and Art Galleries of the Northern Territory); Craig Reid (Queen Victoria Museum & Art Gallery); Genefor Walker-Smith, Kirrily Moore (Tasmanian Museum and Art Gallery); Chris Burwell, Susan Wright, Karin Koch (Queensland Museum); Terry Houston (Western Australian Museum); Dennis Paulson (USA); Gunther Theischinger (EPS, NSW and his personal collection); Fons Peel (of Dragonflypix); Rosser Garrison (his personal collection and that held in the California State Collection of Arthropods); Vincent Kalkman (his collections made in Australia and held in the Naturalis Biodiversity Center); Reiner Richter (data held in the Atlas of Living Australia [Creative Commons Attribution 3.0 Australia (CC BY) Reiner Richter (Odonata), 2 August 2015])

Jim Longworth (Department Agriculture, Water and the Environment) gave invaluable assistance allocating each specimen to its IBRA7 region and Köppen Climate Zone and providing the maps of those regions. The climate maps are covered by the open licence CC - BY. The Bureau of Meteorology is the source of the images.

Overview

This book comprises three sections:

Distribution maps for 325 species of Australian Odonata derived from nearly 60,000 records

Checklists and flight times for each of the 89 Interim Biogeographical Regions of Australia (IBRA7)

Checklists and flight times for each of the 27 Köppen Climate Zones of Australia

Introduction

Sources of Data

Although Australian Institutions now regularly submit their digitized label data to centralised repositories such as the Atlas of Living Australia (ALA) that resource has been used sparingly in the preparation of these distribution maps. They are based primarily on original information from the Collection Managers of the individual museums augmented, in the case of the Australian Museum, with subsequent ALA postings. Individual collectors have also been generous in sharing their specimen data and they are listed in the Acknowledgements section. As each collection has been added to the database the new information has been plotted on the maps and extreme outliers either questioned or removed. There are some which remain but they have been validated. Possibly they are due to vagrant behavior or unusual weather conditions.

Considering those Institutions or Individuals who provide more than 800 records each (and this comprises 96% of the total) we can see their relative contributions in the pie chart below.

Because Australian specimens in their collections are not digitised the substantial holdings at the London Natural History Museum and Naturalis Biodiversity Center, Leiden could not be included.

Acronyms are: ANIC (Australian National Insect Collection, Canberra); QM (Queensland Museum, Brisbane); RNR (Reiner Richter ex Atlas of Living Australia); AM (Australian Musem, Sydney); EPA_ NSW (Environment Protection Authority, NSW); NMV (Museums Victoria, Melbourne); DRP (Dennis Paulson, USA); SAM (South Australian Museum, Adelaide); WAM (Western Australian Museum,

Perth); NMNH (National Museum of Natural History [Smithsonian Institution], Washington D.C.); NTM (Museum and Art Galleries of the Northern Territory, Darwin).

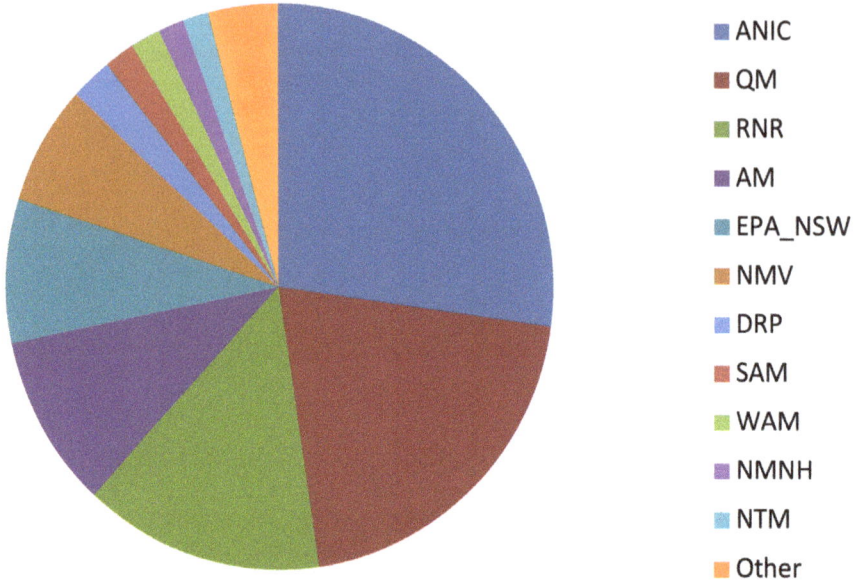

For those who like actual numbers:

ANIC	15,846
QM	11,923
RNR	8,266
AM	5,788
EPA_NSW	4,795
NMV	3,997
DRP	1,448
SAM	1,071
WAM	1,023
NMNH	879
NTM	871
Other	2,390
TOTAL	**58,297**

A project such as this is at the mercy of the idiosyncrasies of the cataloguer. If a specimen exists, its identification can be checked but if erroneous locality information has been given it can lead to large errors. One serious problem is when a specimen label says 'Australia' or even 'New Holland' or even just the State where the specimen was collected and the cataloguer assigns the geographic centroid of Australia or the State as the locality. In reality the maps show the distribution of the activity of the collectors, not insects but, it is hoped with almost 60,000 records, that the distributions are well delineated. It is possibly more of a problem when assessing flight times.

Hierarchy of reliability

The maps are colour-coded to show if the point represents a specimen, literature record, photo or observation. In decreasing order of reliability these categorizations are:

Adult Specimen

> If a properly curated specimen is available its identification can be confirmed. There can be some doubt, occasionally, about locality information. Illegible writing on a label or different localities with the same name can lead to errors. Extreme outliers on the maps have been removed.

Literature

> There are relatively few instances in the database and, generally, they are from the original description, therefore correct by definition.

Larval Specimen

> Larvae are more difficult to identify than adults and for some no morphologically distinguishing characteristics have yet been found.

> Obviously, they can't contribute to flight-time phenology

Photo

> Even good photos do not always show in sufficient detail characters which are essential to separate some very similar species. Tenerals and some females are particularly problematic.

Observation

> This depends on the skills and experience of the observer and, for many species, it is the most unreliable method of recording presence. With some observers it also suffers from the desire to spot rarities.

This chart shows that the bulk of the records come from adult dragonflies thus giving a high degree of confidence in the maps.

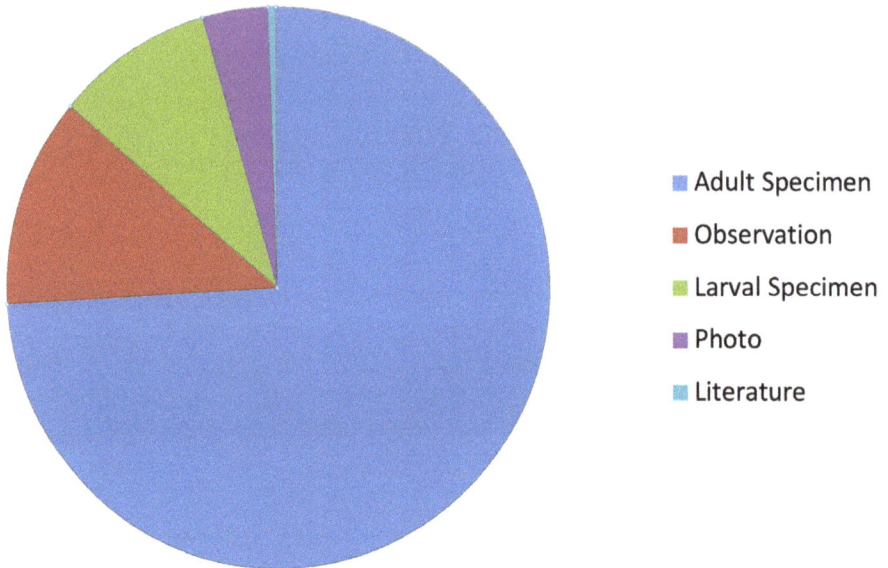

- Adult Specimen
- Observation
- Larval Specimen
- Photo
- Literature

Precision of Maps and Observations

Each map square covers an area of 30' longitude x 30' latitude. On average, in Australia, this represents an area 55 km (N-S) x 45 km (E-W).

Degrees of latitude are parallel so the distance between each degree remains almost constant but since degrees of longitude are farthest apart at the equator and converge at the poles, their distance varies greatly.

Each degree of latitude is approximately 111 kilometers apart. Due to the earth's slightly ellipsoid shape the range varies from 110.567 km at the equator to 111.699 km at the poles. A degree of longitude is widest at the equator at 111.321km and gradually lessens to zero at the poles. The length of 1 degree of Longitude = cosine (latitude in decimal degrees) * length of degree at equator. At the tip of Cape York, Australia's northernmost point [10° 41' 14.09"S, 142° 31' 52.88 E] one degree of longitude is 109 km while at the southernmost tip of Tasmania [43° 39' 14.73"S, 146° 51' 01.28 E] it is 89 km.

A change of 1 in the third decimal place of a decimal degree means only about 90 metres on the ground. Quoting six decimal points (= 0.003") (as some do) implies an accuracy of better than 10 centimetres.

What the Maps Tell Us

The database from which these maps were produced contained 58, 297 separate entries, each representing a curated specimen, a literature record, a photograph or a sighting with a high probability of being correctly identified. Dbase, the program used to generate the maps only marks one square for each collection event (i.e. species, locality, date) and it has 35,900 entries. So, 38% of records came from multiple captures from the same locality on the same date.

Four of those specimens are *Agriocnemis exsudans* from the Australian Territory of Norfolk Island and are not included on the maps of the mainland. Twelve specimens have no coordinates but do have a locality given which might enable them to be mapped if addressed by someone with local knowledge.

There are maps for 325 species. Half the species account for 94% of the database occurrences. It is a very skewed distribution. Invoking Pareto's Rule we see that 80% of the specimens are from 87 species, 27% of the total number.

Fifty-three species have less than 10 occurrences in the database while five have only one:

Austrogomphus pusillus Sjöstedt, 1917; *Eusynthemis cooloola* Theischinger, 2018; *Hemigomphus atratus* Watson, 1991; *Micromidia rodericki* Fraser, 1959; *Telephlebia undia* Theischinger, 1985.

Eusynthemis cooloola has been recently described from a single adult male specimen collected in 1984.

The top 10, each with more than 1,000 specimens, photos or sightings are:

> *Ischnura heterosticta*
>
> *Diplacodes bipunctata*

Austroargiolestes icteromelas

Orthetrum caledonicum

Hemicordulia tau

Ischnura aurora

Diplacodes haematodes

Hemianax papuensis

Xanthagrion erythroneurum

Austrolestes leda

How many entries make good map? If we assume that number to be 30 (*cf.* maps for *Episynlestes cristatus* or *Petalura ingentissima*) then only 221 qualify; 32% are inadequate. If 20 points are considered enough (*cf* maps for *Austrogomphus arbustorum* or *Zyxomma petiolatum*) then another 18 would be added but the number of inadequate ones is still slightly more than a quarter. The geographic spread of the points has quite a bearing on these appearances.

Flight Seasons

An important aspect of the phenology of Dragonflies is the studying the flight times of Adults. Date of capture might be a possible surrogate to analyse flight seasons. However there are some difficulties:

1. Australia has a large latitudinal and climate range;

2. Actually we are sampling the visits of collectors rather than dragonflies;

3. Some collectors take a large series of specimens; others don't. This could bias the sampling and analysis.

Ideally, the sampling effort should have been consistent across all months within a specified zone, using the same protocol.

Zonation

Some dragonflies have a large latitudinal range within Australia. Comparing the flight times of *Hemianax papuensis* from Cape York with those from central Tasmania bears no sense. Similarly for *Ischnura aurora, Ischnura heterosticta* and any other species with a large geographic range. One solution to this might be to consider partitioning the lists into discrete regions. Based on global ecoregions the Interim Biogeographic Regionalisation for

Australia (IBRA), was prepared giving distinct bioregions based on common climate, geology, landform, native vegetation and species information. IBRA was developed in 1993-94 and version 6.1 was published in 2004. It comprised 85 bioregions and 403 subregions, subregions are more localised and homogenous geomorphological units in each bioregion. Descriptions of each bioregion were taken from internet listings for IBRA6.1. In 2014 IBRA 7 defined 89 bioregions and 419 subregions, adding four new oceanic bioregions: the Indian Tropical Islands Bioregion, the Pacific Subtropical Islands Bioregion, the Subantarctic Islands Bioregion and the Coral Sea Bioregion. These bioregions account for Australia's island territories including Christmas Island in the Indian Ocean, Macquarie Island in the Southern Ocean, Norfolk and Lord Howe Islands in the Pacific Ocean and the Coral Sea Islands Territory. Flight time charts, which also serve as local checklists, for the IBRA 7 zones can be found between pages 69 and 241. Almost 10% of records are from larvae but collection dates of specimens of larvae cannot contribute to flight calendars for adults so they have been ignored. However, on the rare occasion where a larva species has been sampled without a comparable adult being recorded, that larva is included in the total number of species.

The totals in the flight charts are raw numbers of specimens or observations. No attempt has been made to standardize (or colour-code) them for comparison. In this way they give an appreciation of the total numbers that have been included.

In hindsight a zonation based on geomorphological and vegetation characteristics, such as the IBRA parameters is not consistent with dragonfly biology. Hydrology and temperature are likely to be more relevant so perhaps climate zones should be used. The **Köppen climate classification** is one of the most widely used climate classification systems. It was first published by the German-Russian climatologist Wladimir Köppen from the University of Graz, Austria in 1884. It is based on the assumption that native vegetation is the best indicator of climate zone boundaries. The classification divides climates into five main climate groups, with each group, and its subgroups, being based on seasonal precipitation and temperature patterns. His five main groups are *A* (tropical), *B* (dry), *C* (temperate), *D* (continental), and *E* (polar). Stern *et al* (2000)[4] modified the scheme for Australian conditions, using mean monthly rainfall, mean annual rainfall, mean maximum temperature and mean minimum temperature from the Bureau of Meteorology's records for the 30-year period 1961-1990. The resulting 27 zones, with the number of records and number of species of dragonflies, are:

---

4. Stern H.,de Hoedt, G. & Ernst, J. (2000). Objective classification of Australian climates. *Australian Meteorological Magazine* 49: 87-96

	Species	Specimens
Desert - Hot (persistently dry)	28	448
Desert - Hot (Summer drought)	12	89
Desert - Hot (Winter drought)	16	67
Desert - Warm (persistently dry)	9	21
Equatorial - Rainforest (monsoonal)	12	27
Equatorial - Savanna	105	2423
Grassland - Hot (persistently dry)	52	1463
Grassland - Hot (Summer drought)	41	578
Grassland - Hot (Winter drought)	91	1421
Grassland - Warm (persistently dry)	38	1204
Grassland - Warm (Summer drought)	5	10
Subtropical - Distinctly dry summer	35	1086
Subtropical - Distinctly dry winter	106	1111
Subtropical - Moderately dry winter	108	1616
Subtropical - No dry season	190	10251
Temperate - Distinctly dry (and hot) Summer	40	925
Temperate - Distinctly dry (and mild) Summer	1	1
Temperate - Distinctly dry (and warm) Summer	54	1387
Temperate - Moderately dry Winter (hot Summer)	43	202
Temperate - Moderately dry Winter (warm Summer)	11	15
Temperate - No dry season (cool Summer)	23	206
Temperate - No dry season (hot Summer)	123	4482
Temperate - No dry season (mild Summer)	95	4788
Temperate - No dry season (warm Summer)	126	16214
Tropical - Rainforest (monsoonal)	127	2341
Tropical - Rainforest (persistently wet)	103	1071
Tropical - Savanna	155	4828

Flight time charts, which also serve as local checklists, for the Australian Köppen climate zones can be found between pages 242 and 318. The same caveats which relate to the use of larvae in IBRA7 zones also apply here.

It could be constructive to see whether there are any common factors in the most populous climate zones that determine dragonfly preference. Eighty percent of the specimens come from just eight climate zones.

These are, in descending order:

- Temperate - No dry season (warm Summer)

- Subtropical - No dry season

- Tropical - Savanna

- Temperate - No dry season (mild Summer)

- Temperate - No dry season (hot Summer)

- Equatorial - Savanna

- Tropical - Rainforest (monsoonal)

- Subtropical - Moderately dry winter

There seems to be no common factor standing out from these groups except No Dry Season which is understandable.

Inequitable Sampling

The problem of inequitable sampling is best demonstrated by reference to IBRA7 zone DEU – Desert Uplands – an area in central Queensland.

Desert Uplands

While there are records of 140 specimens and 32 species they were made over only effectively five visits most of which were in April.

16-April-1973
30-April-1980
08-April, 09-April, 11-April, 12-April, 13-April, 14-April, 15-April, 16-April, 17-April, 18-April-2000
16-December-2006
28-October-2010

Checklists

The checklists which are the basis for the individual flight time charts are based on:

The classification and diversity of dragonflies and damselflies (Odonata) (2013) Dijkstra, K.-D. B., Bechly, G., Bybee, S.M., Dow, R.A., Dumont, H.J., Fleck, G., Garrison, R.W., Hämäläinen. M., Kalkman, V.J., Karube, H., May, M.L., Orr, A.G., Paulson, D.R., Rehn,

A.C., Theischinger, G., Trueman, J.W.H., van Tol, J., von Ellenreider, N. & Ware, J. pp. 36-45 in: Zhang, Z-Q. (Ed.) Animal Diversity: An Outline of Higher Classification and Survey of Taxonomic Richness (Addenda). *Zootaxa* 3703: 1-82.

Generic revision of Argiolestidae (Odonata), with four new genera. (2013) Kalkman, V.J & Theischinger, G. *International Journal of Odonatology* 16(1): 1-52

The authors of the first article consider the Australian genera *Apocordulia*, *Archaeophya*, *Austrocordulia*, *Austrophya*, *Cordulephya*, *Hesperocordulia*, *Lathrocordulia*, *Micromidia*, and *Pseudocordulia* to be *incertae sedis* within the Libelluloidea.

Austrocoenagrion [as opposed to *Coenagrion lyelli*] was resurrected in Dijkstra, K.-D.B., & Kalkman, V.J. 2012. Phylogeny, classification and taxonomy of European dragonflies and damselflies (Odonata): a review. *Organisms Diversity and Evolution* 12: 209-227, and repeated in Dijkstra, K.-D.B., Kalkman, V.J., Dow, R.A., Stokvis, F.R. & van Tol, J. 2013. Redefining the damselfly families: a comprehensive molecular phylogeny of Zygoptera (Odonata). *Systematic Entomology* 39: 68-96.

Distribution Maps

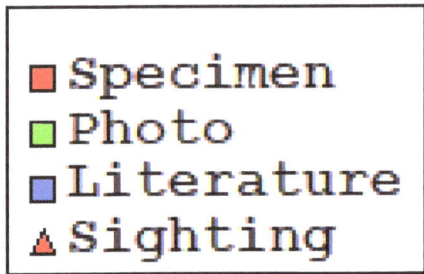

- ■ Specimen
- ■ Photo
- ■ Literature
- ▲ Sighting

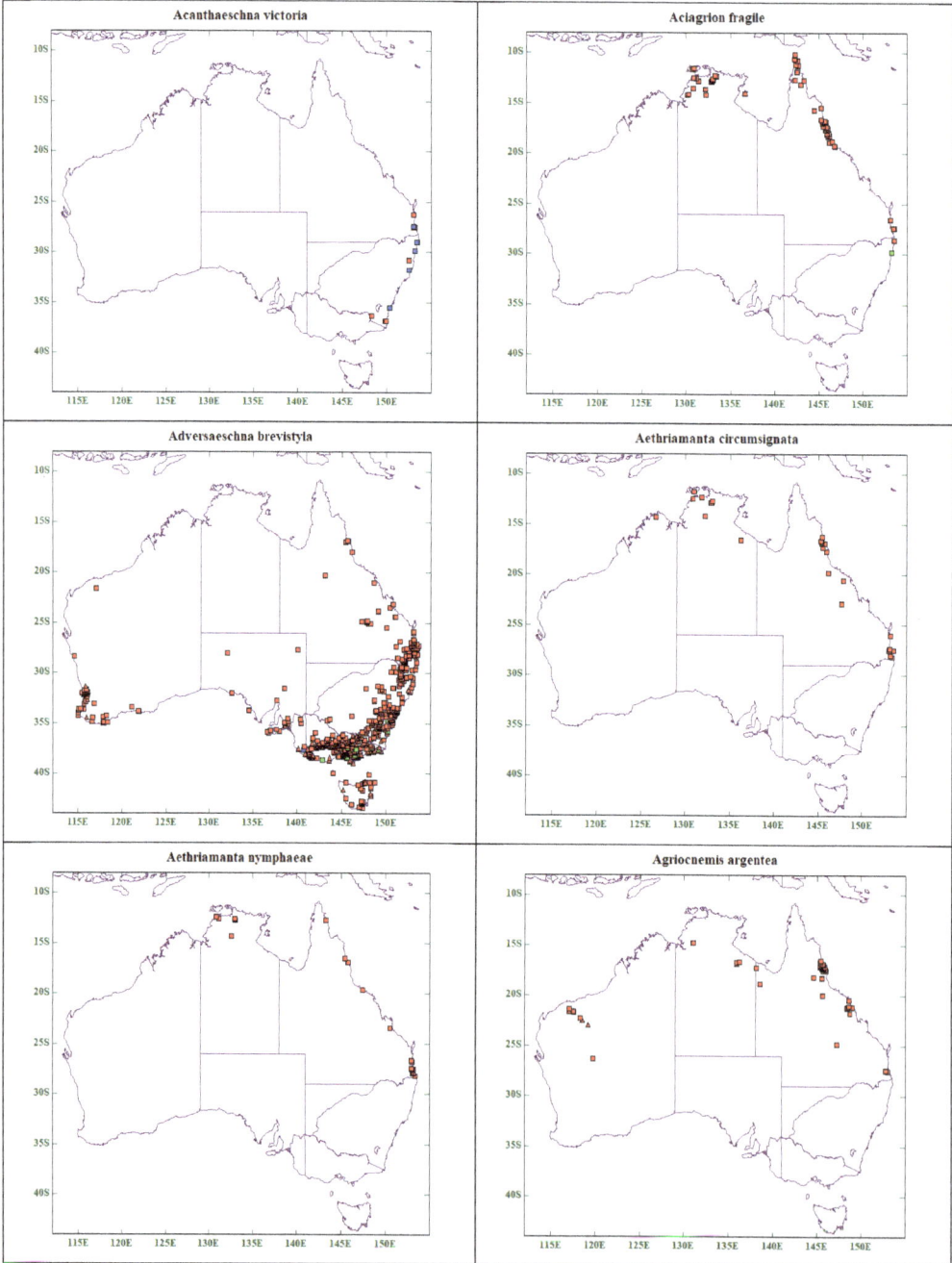

Acanthaeschna victoria

Aciagrion fragile

Adversaeschna brevistyla

Aethriamanta circumsignata

Aethriamanta nymphaeae

Agriocnemis argentea

Agriocnemis dobsoni

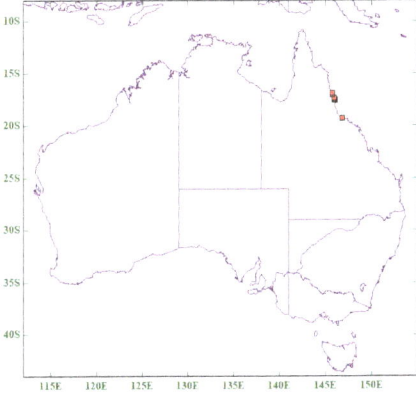

Agriocnemis femina

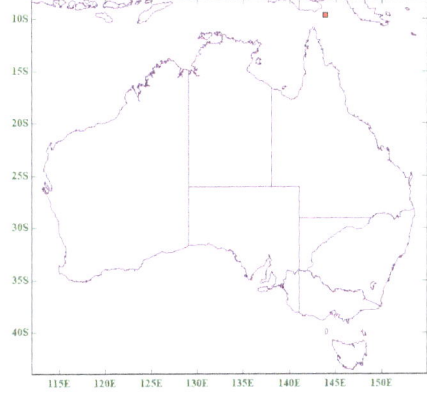

Agriocnemis kunjina

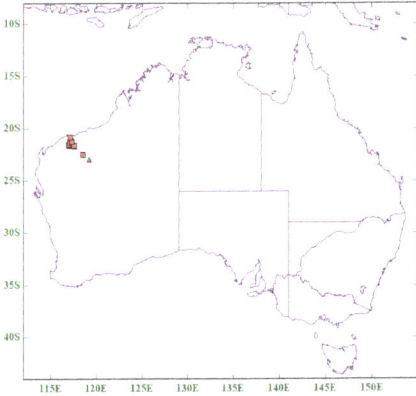

Agriocnemis pygmaea

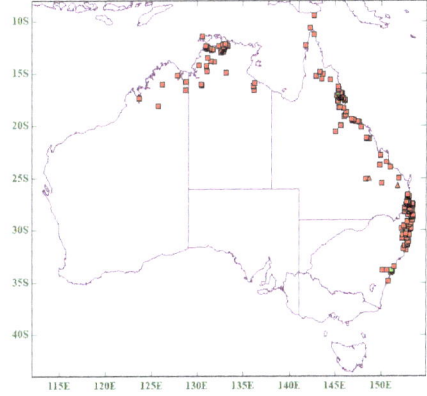

Agriocnemis rubricauda

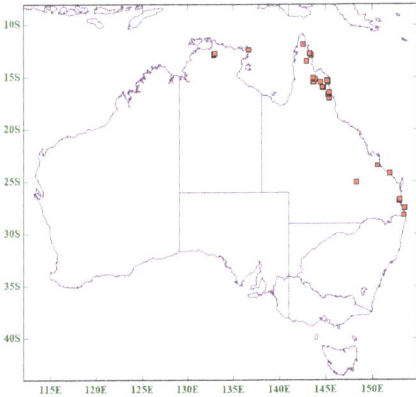

Agrionoptera insignis allogenes

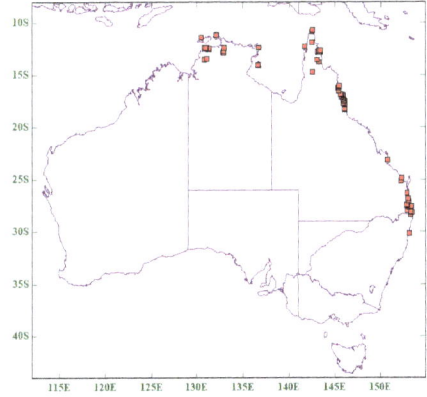

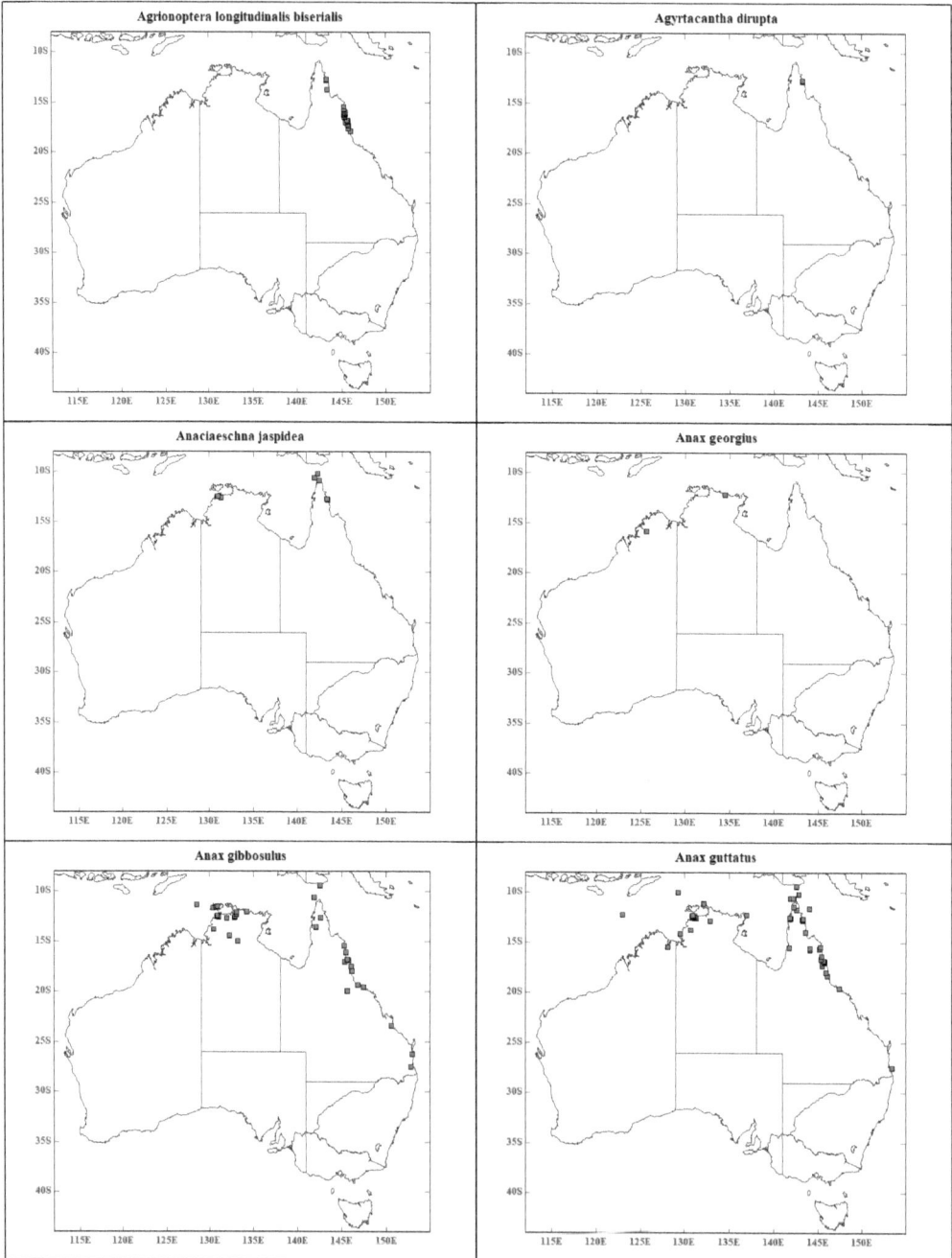

Agrionoptera longitudinalis biserialis

Agyrtacantha dirupta

Anaciaeschna jaspidea

Anax georgius

Anax gibbosulus

Anax guttatus

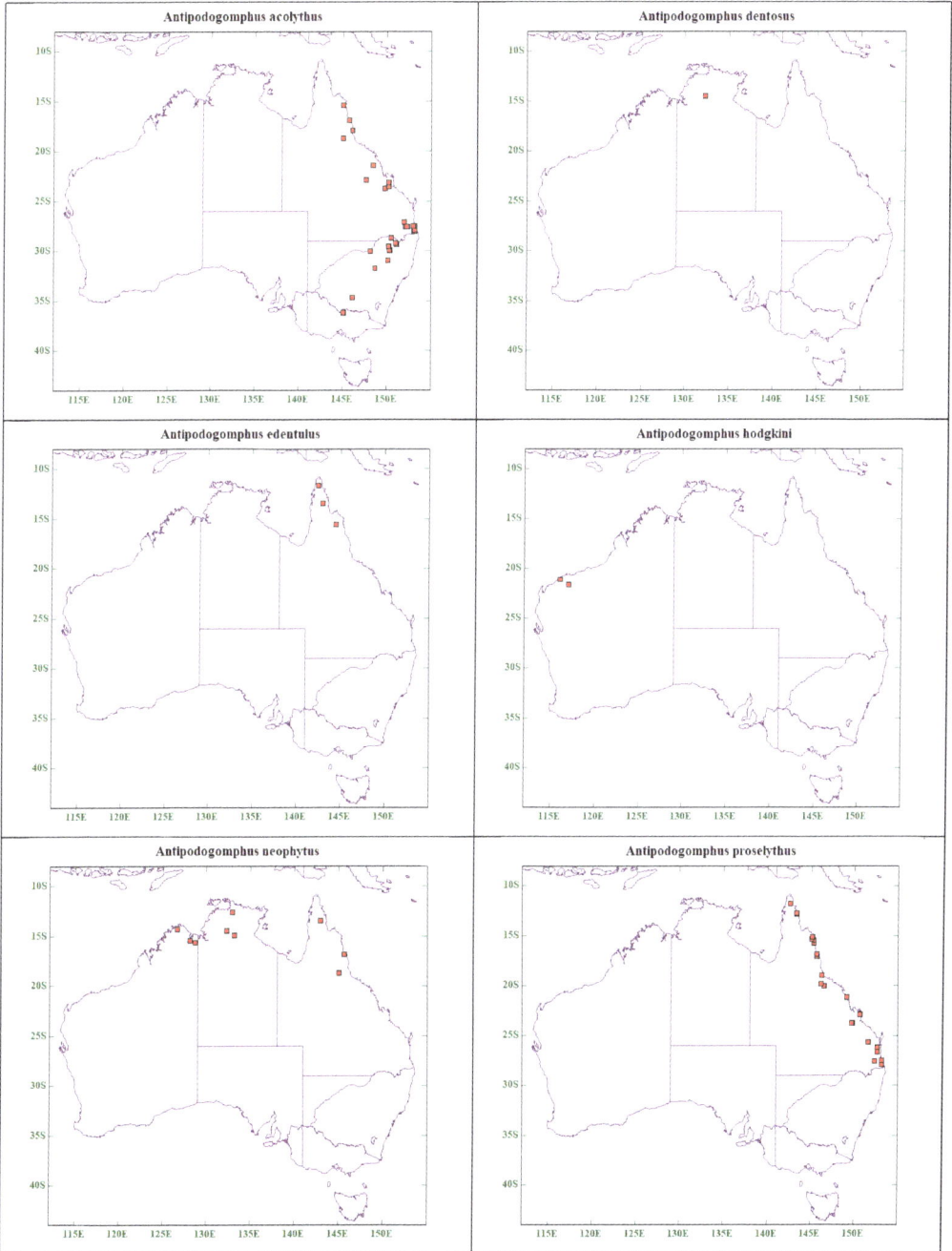

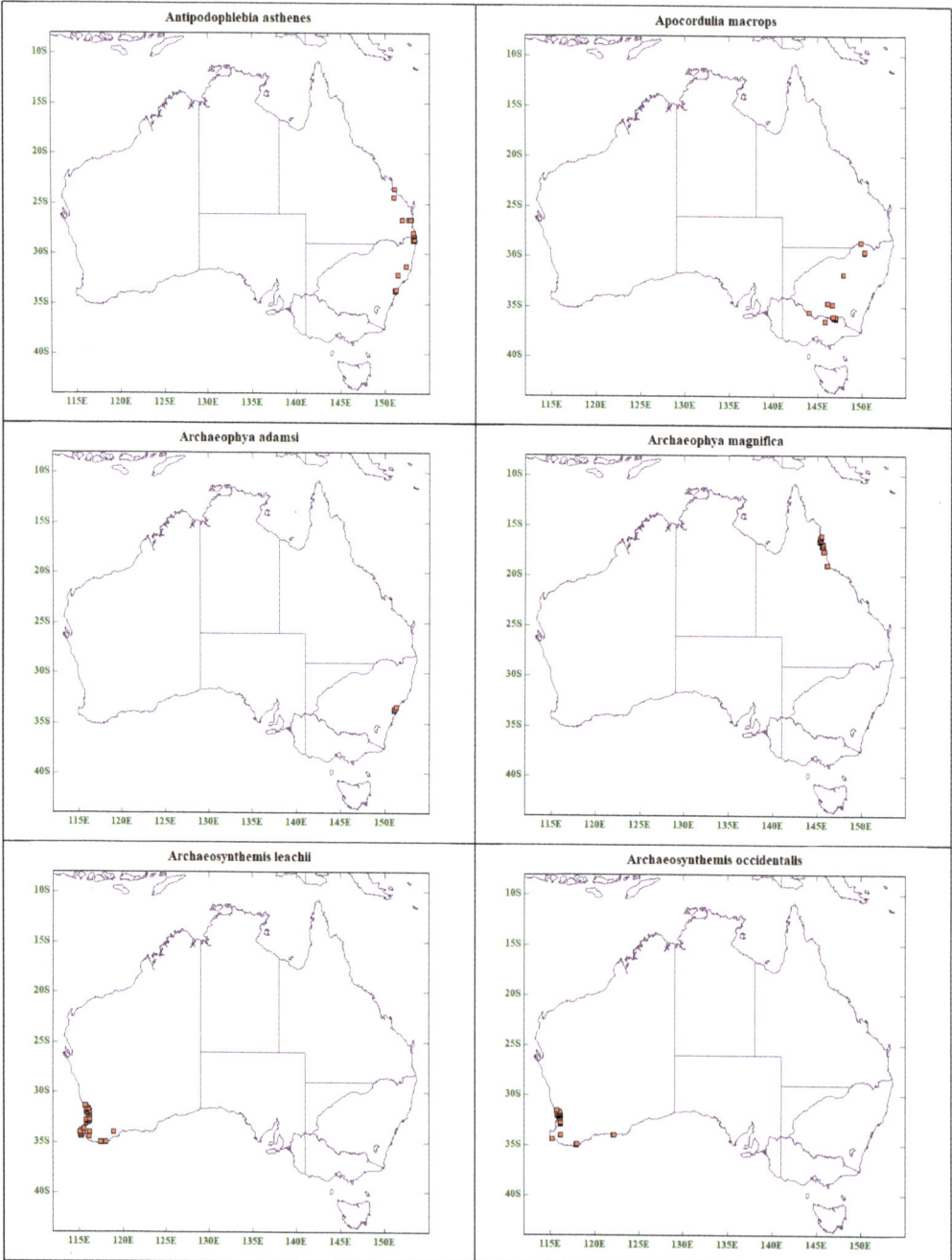

Antipodophlebia asthenes

Apocordulia macrops

Archaeophya adamsi

Archaeophya magnifica

Archaeosynthemis leachii

Archaeosynthemis occidentalis

Archaeosynthemis orientalis

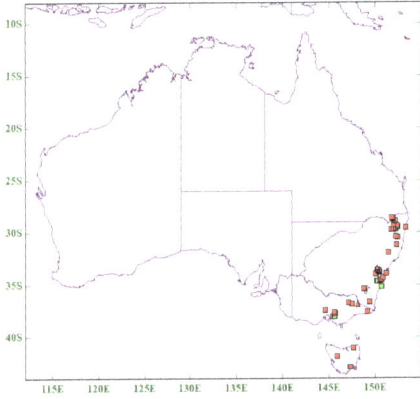

Archaeosynthemis spiniger

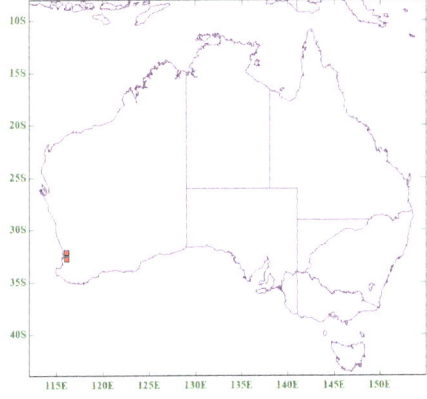

Archiargiolestes parvulus

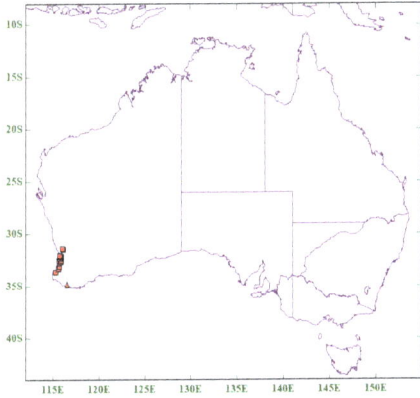

Archiargiolestes pusillissimus

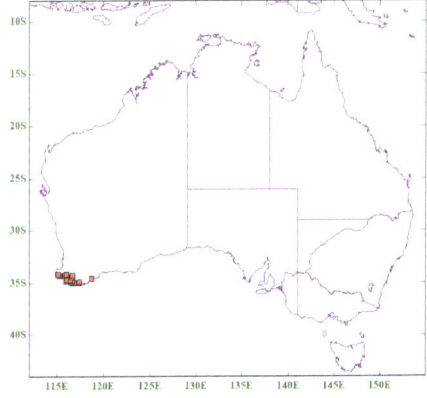

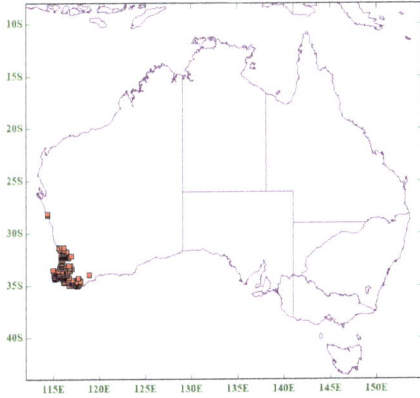

Archiargiolestes pusillus

Archibasis mimetes

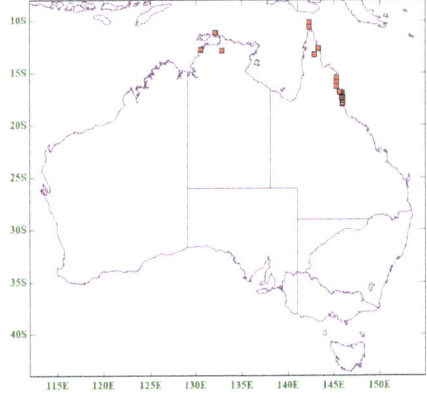

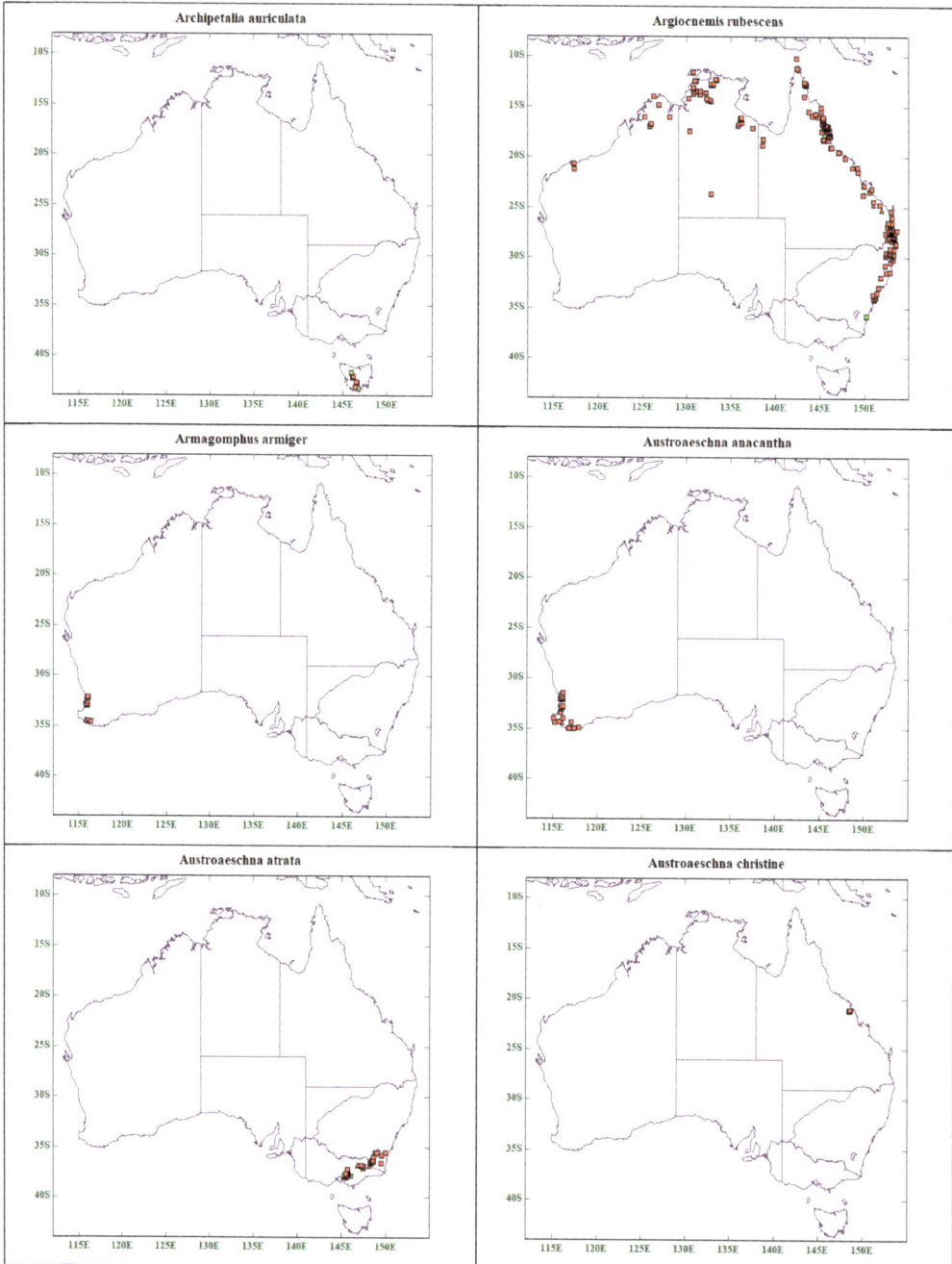

Archipetalia auriculata

Argiocnemis rubescens

Armagomphus armiger

Austroaeschna anacantha

Austroaeschna atrata

Austroaeschna christine

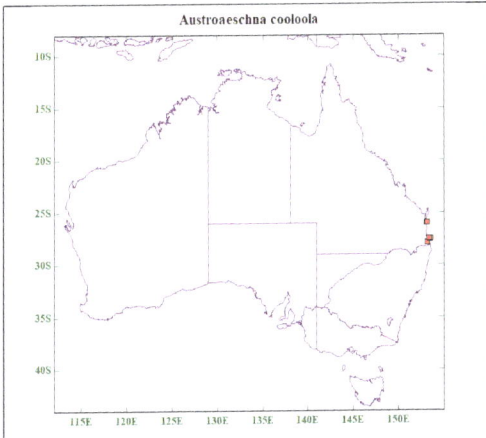

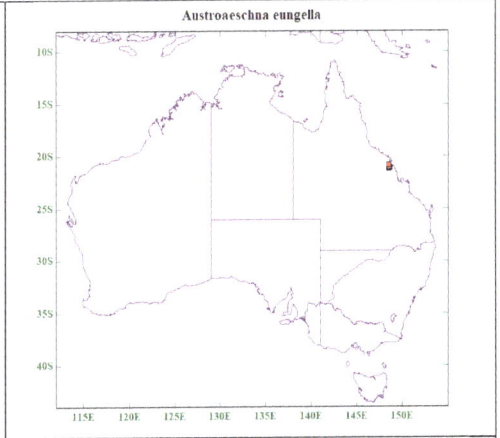

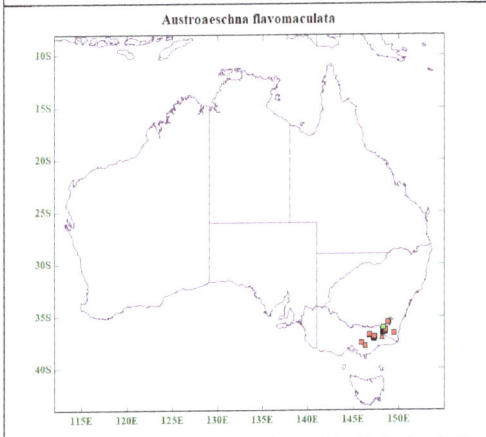

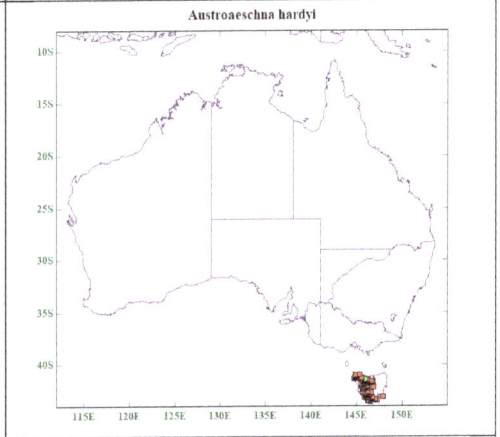

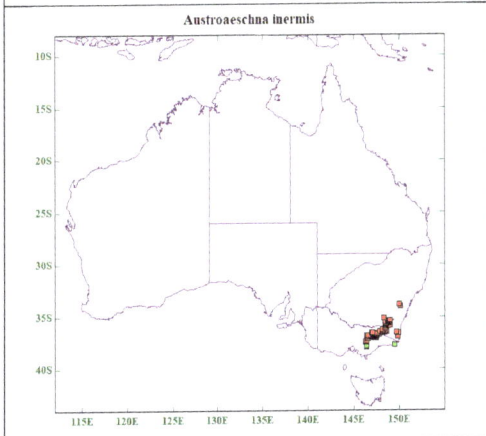

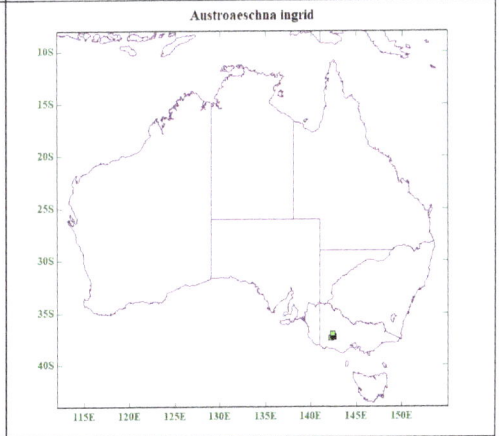

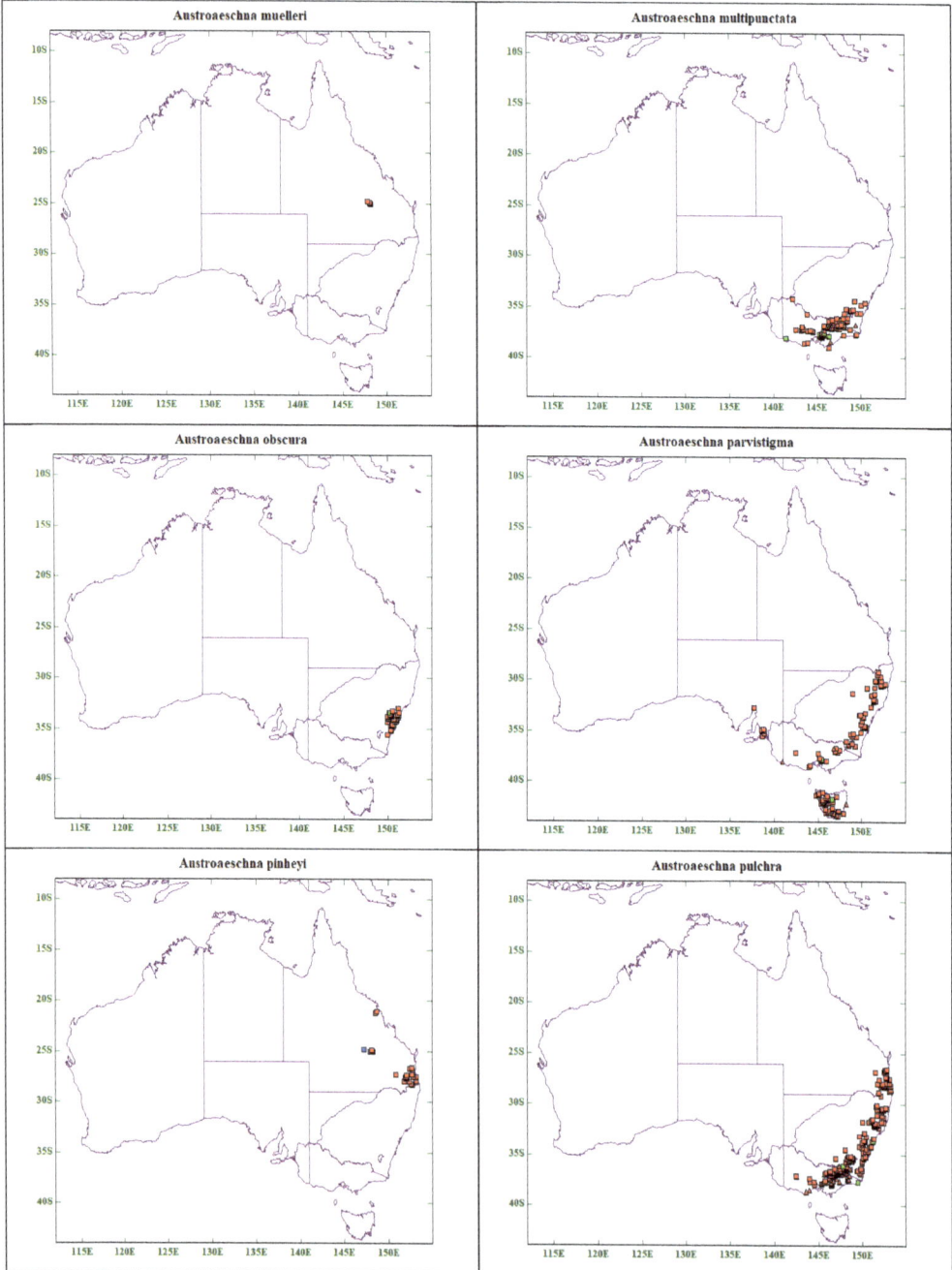

Austroaeschna muelleri

Austroaeschna multipunctata

Austroaeschna obscura

Austroaeschna parvistigma

Austroaeschna pinheyi

Austroaeschna pulchra

Austroaeschna sigma

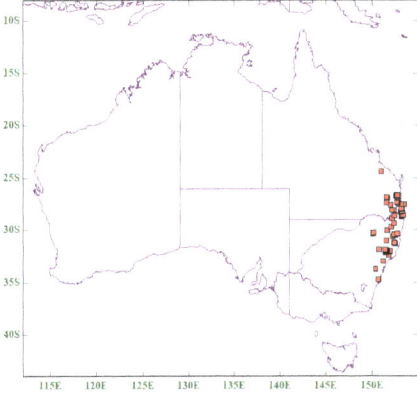

Austroaeschna speciosa

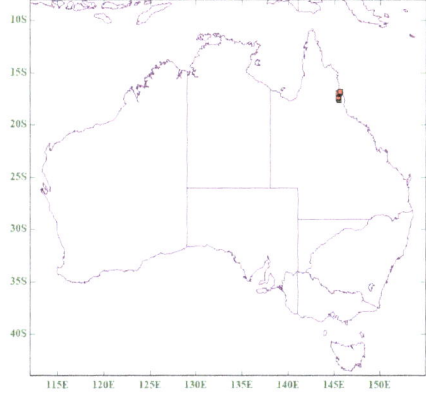

Austroaeschna subapicalis

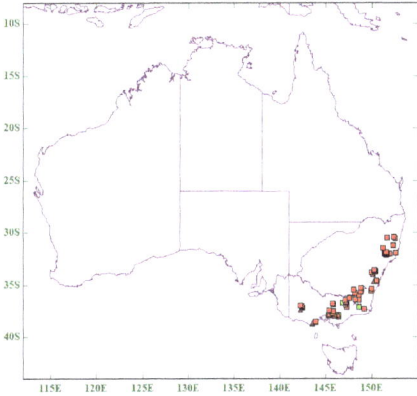

Austroaeschna tasmanica

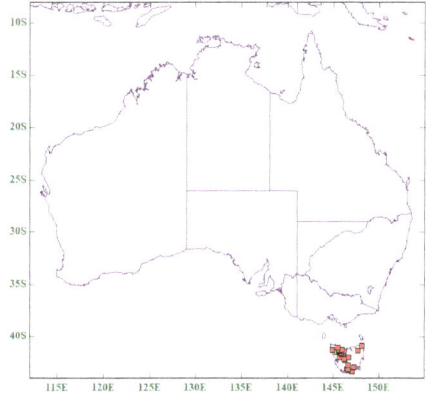

Austroaeschna unicornis

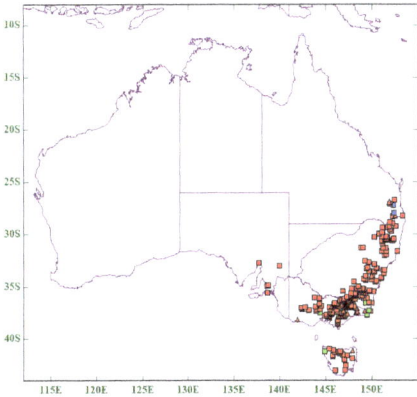

Austroagrion cyane

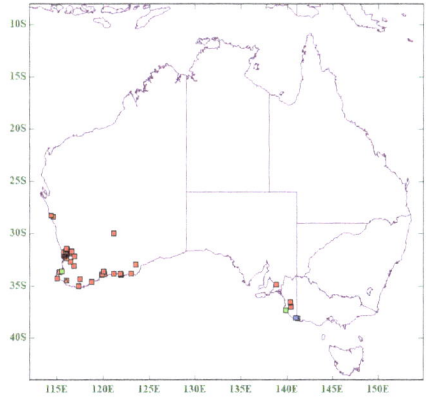

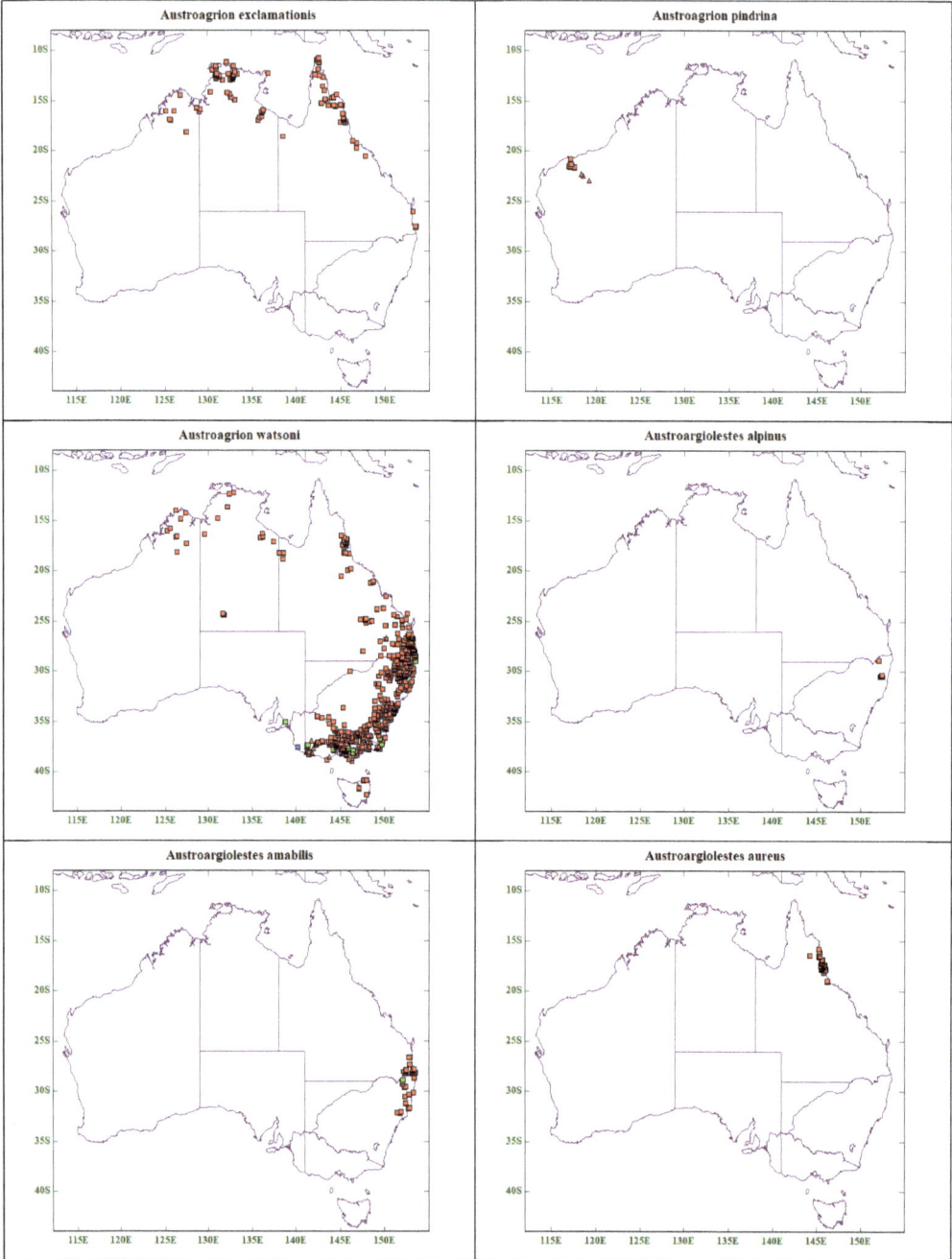

Austroagrion exclamationis

Austroagrion pindrina

Austroagrion watsoni

Austroargiolestes alpinus

Austroargiolestes amabilis

Austroargiolestes aureus

Austroargiolestes brookhousei

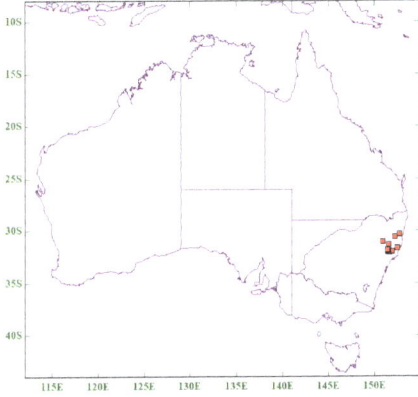

Austroargiolestes calcaris

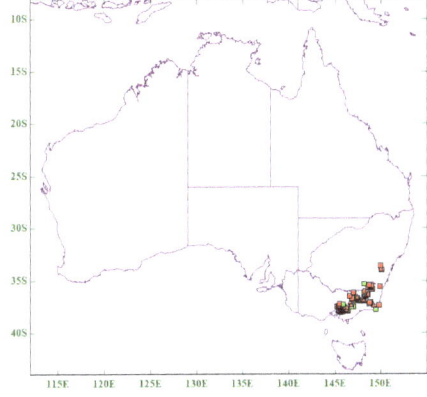

Austroargiolestes christine

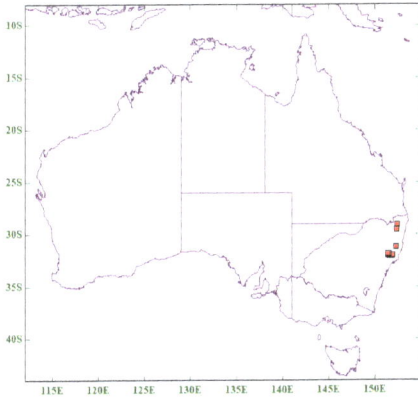

Austroargiolestes chrysoides

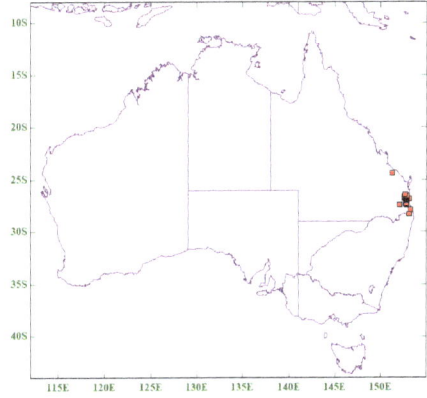

Austroargiolestes elke

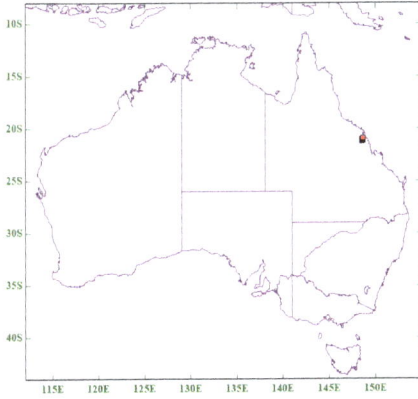

Austroargiolestes icteromelas

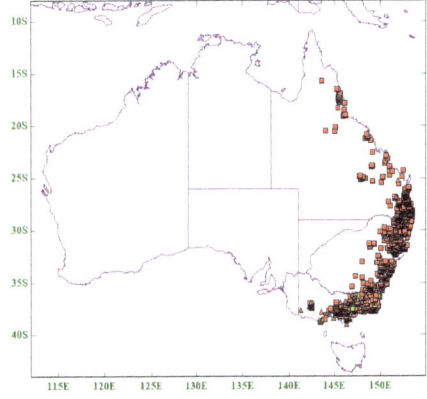

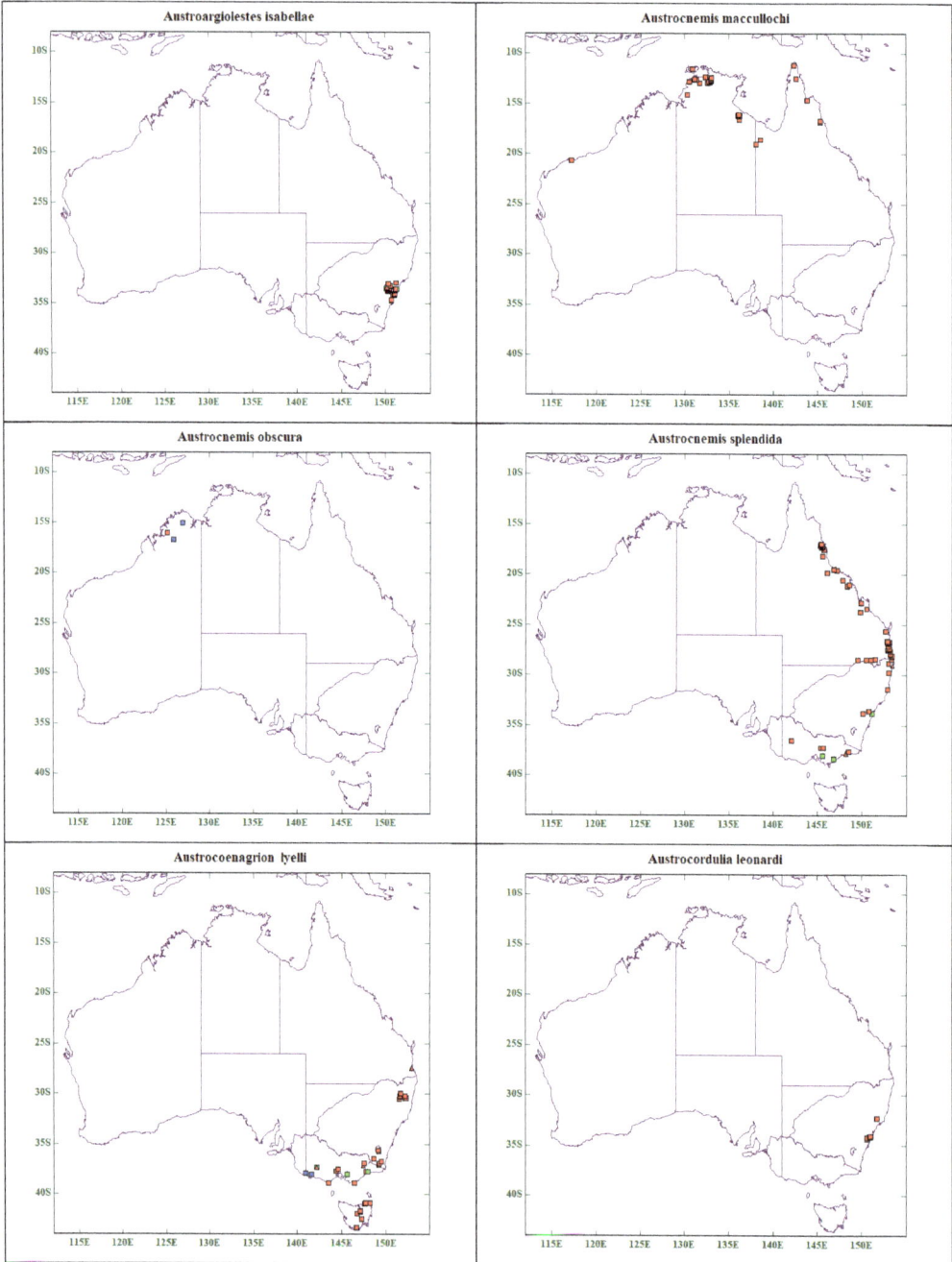

Austroargiolestes isabellae

Austrocnemis maccullochi

Austrocnemis obscura

Austrocnemis splendida

Austrocoenagrion lyelli

Austrocordulia leonardi

Austrocordulia refracta

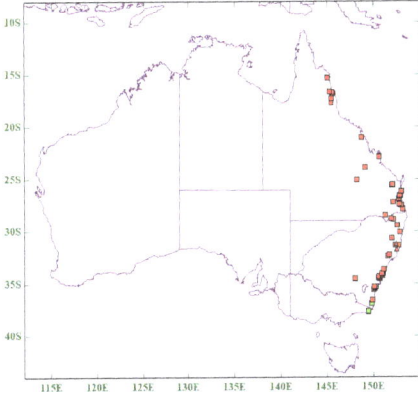

Austrocordulia territoria

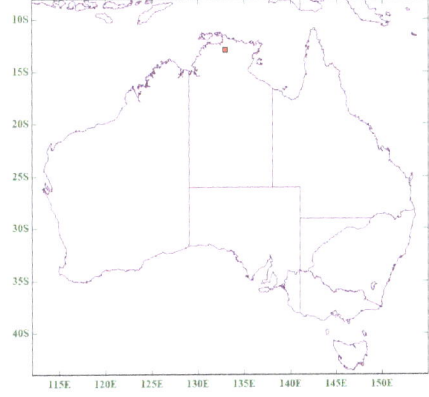

Austroepigomphus gordoni

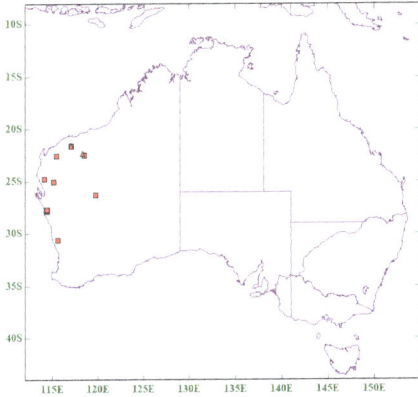

Austroepigomphus praeruptus

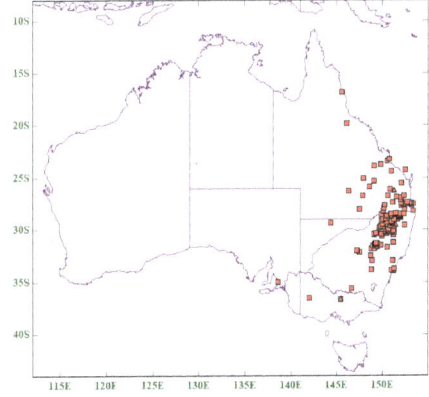

Austroepigomphus turneri

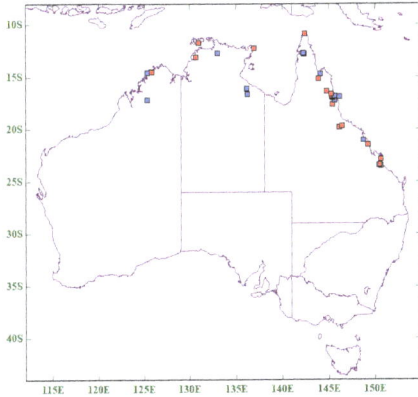

Austrogomphus amphiclitus

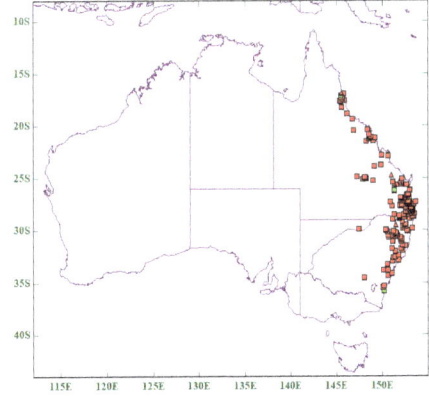

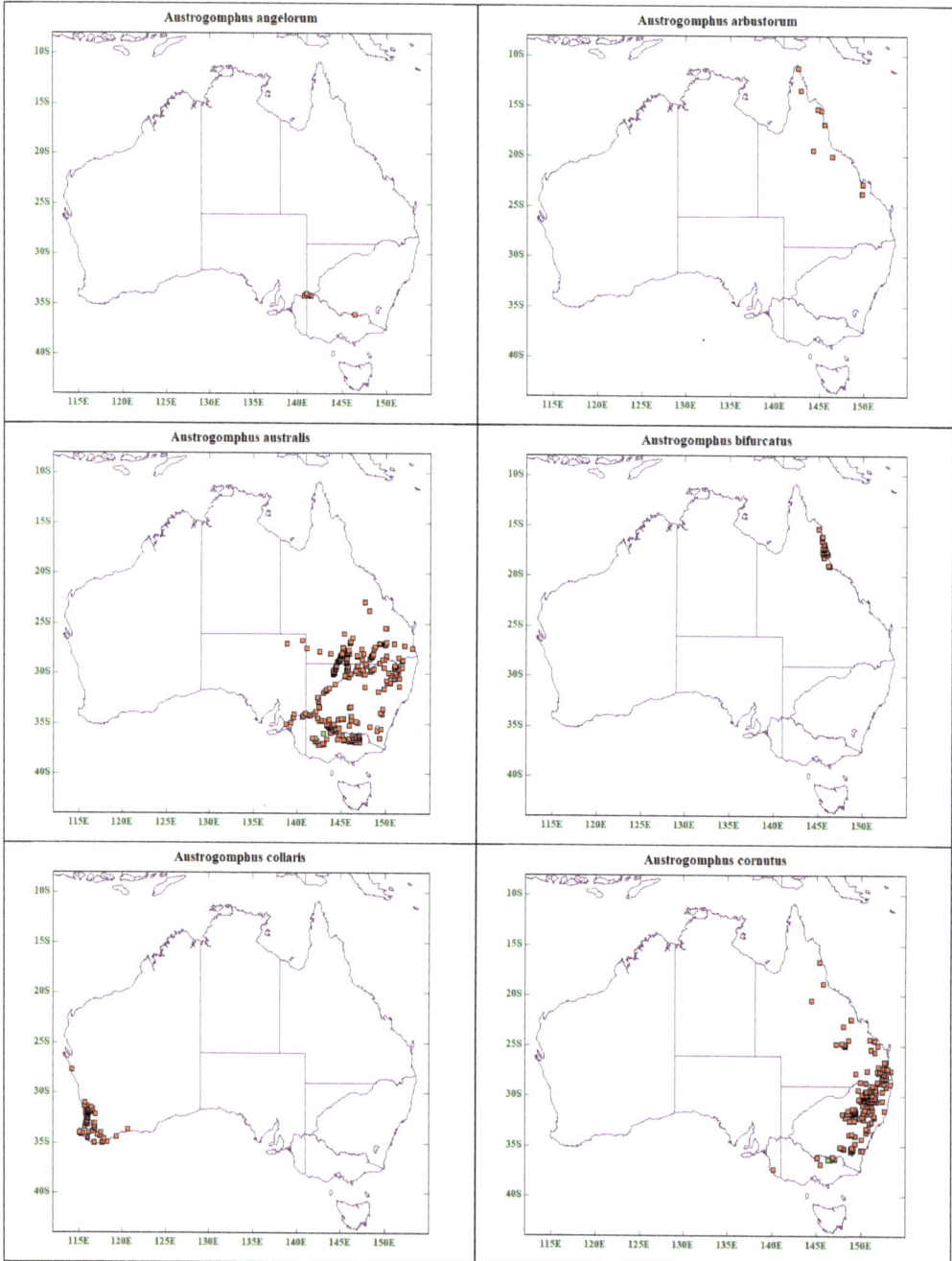

Austrogomphus angelorum

Austrogomphus arbustorum

Austrogomphus australis

Austrogomphus bifurcatus

Austrogomphus collaris

Austrogomphus cornutus

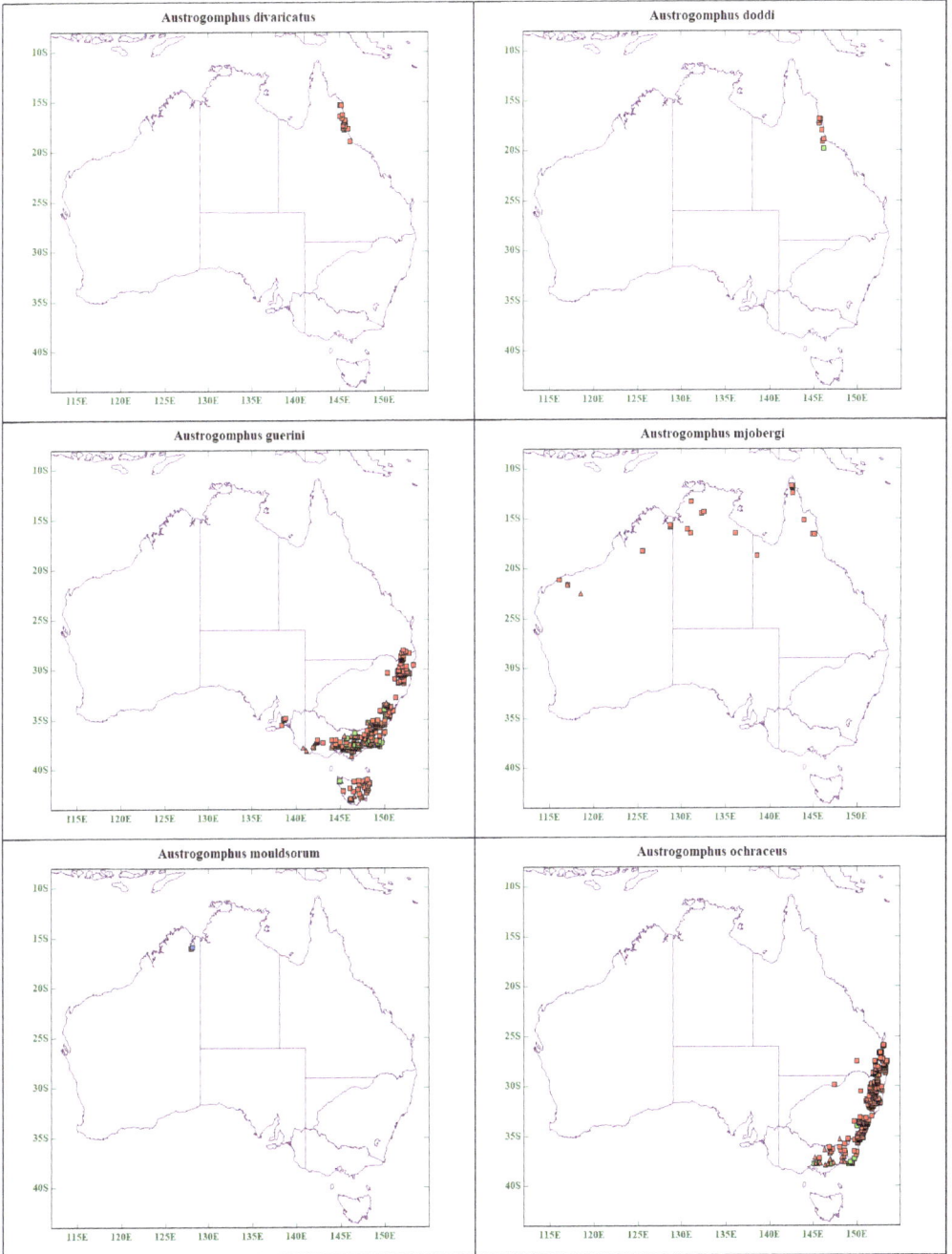

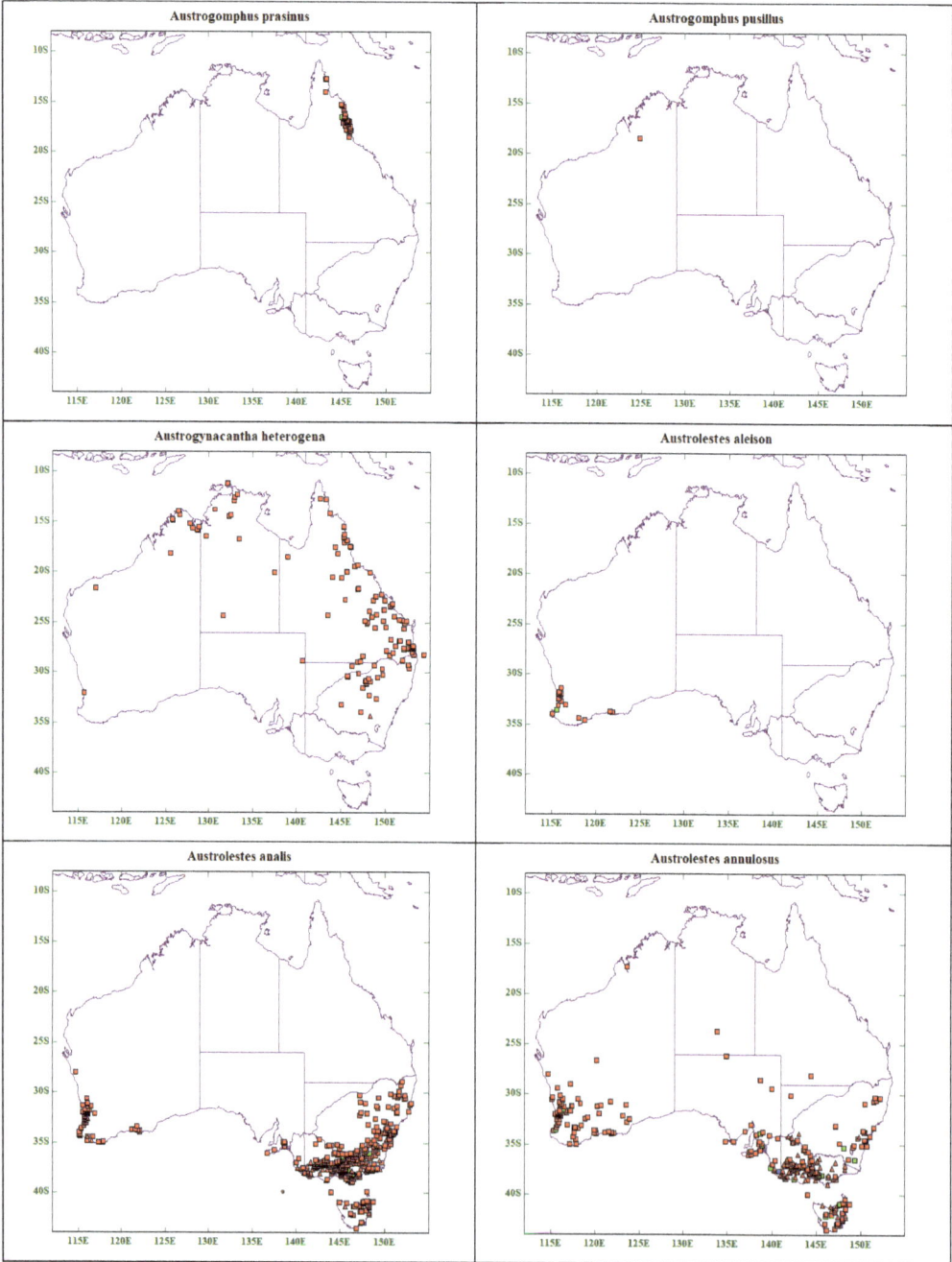

Austrogomphus prasinus

Austrogomphus pusillus

Austrogynacantha heterogena

Austrolestes aleison

Austrolestes analis

Austrolestes annulosus

Austrolestes aridus

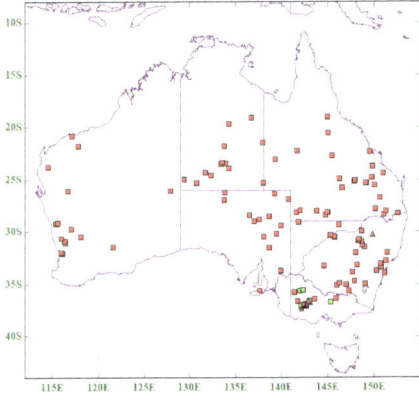

Austrolestes cingulatus

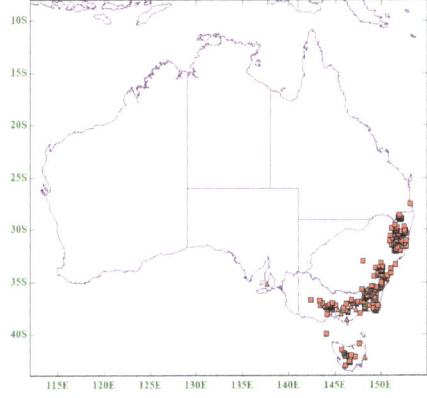

Austrolestes insularis

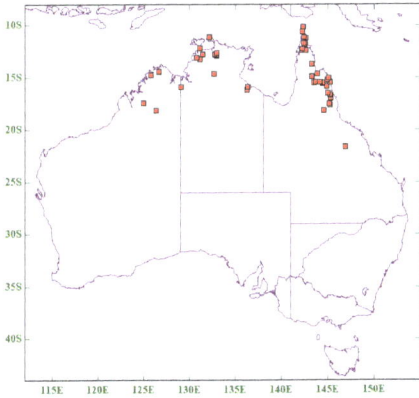

Austrolestes io

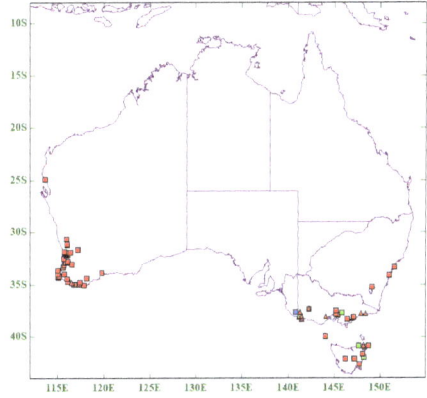

Austrolestes leda

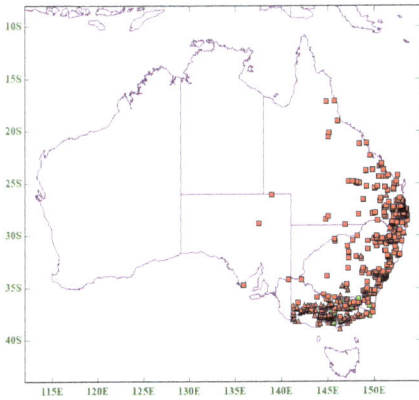

Austrolestes minjerriba

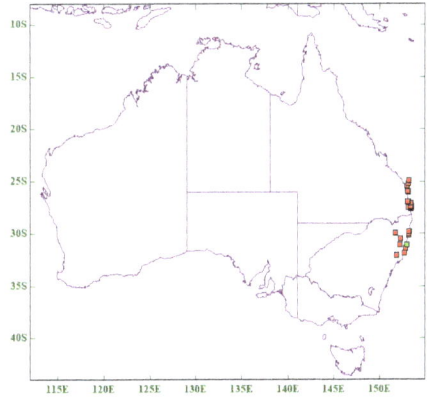

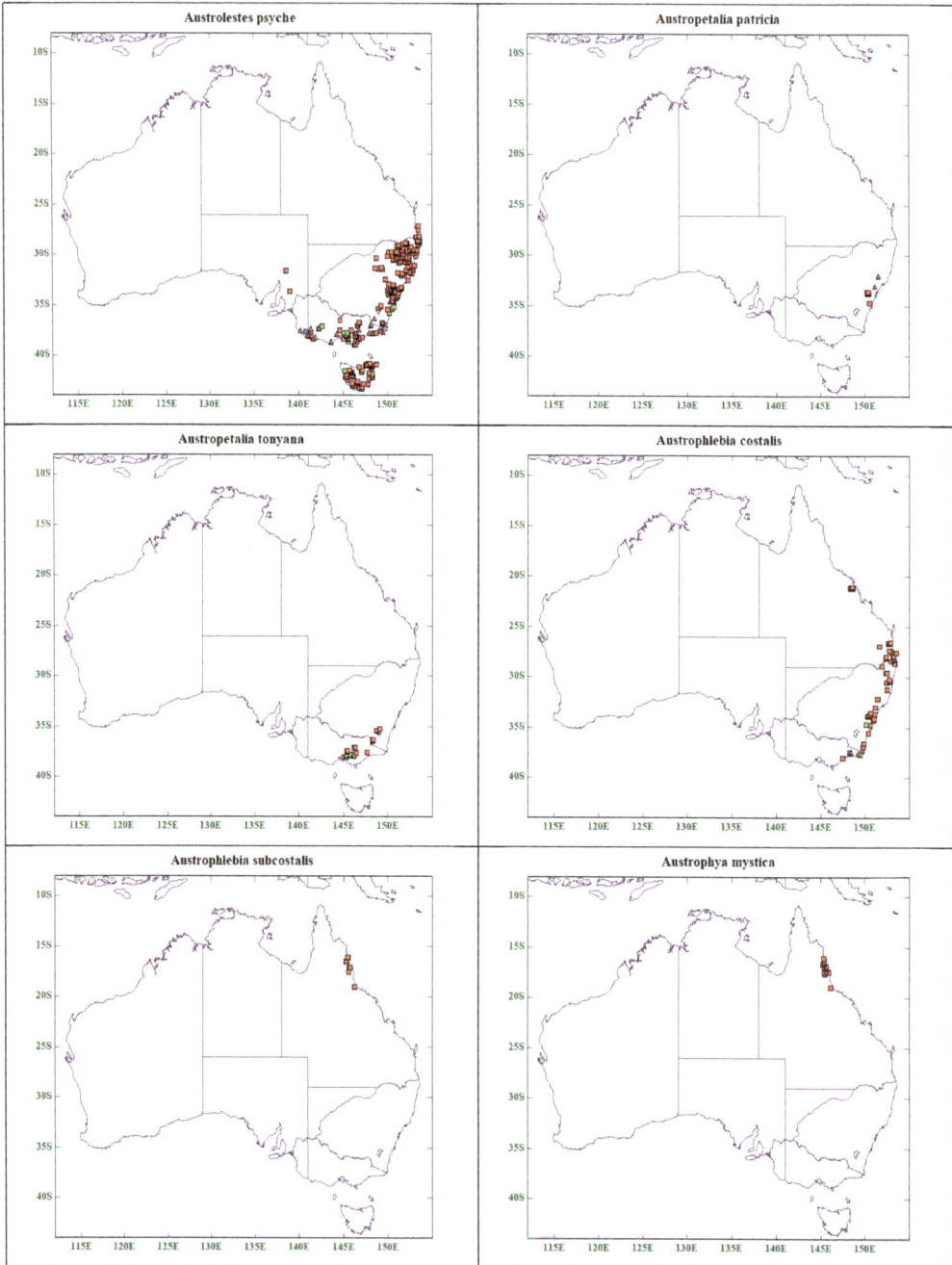

Austrolestes psyche

Austropetalia patricia

Austropetalia tonyana

Austrophlebia costalis

Austrophlebia subcostalis

Austrophya mystica

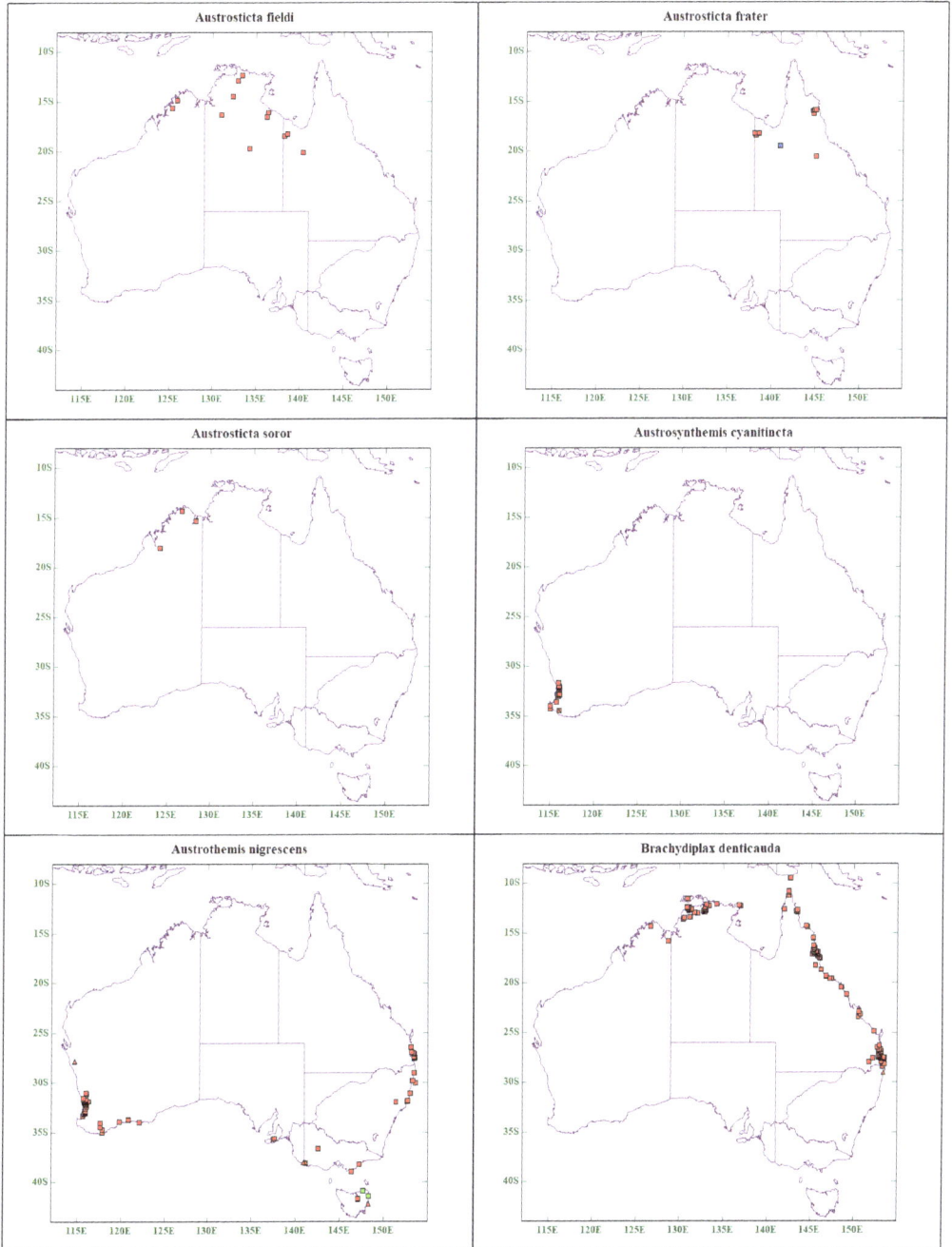

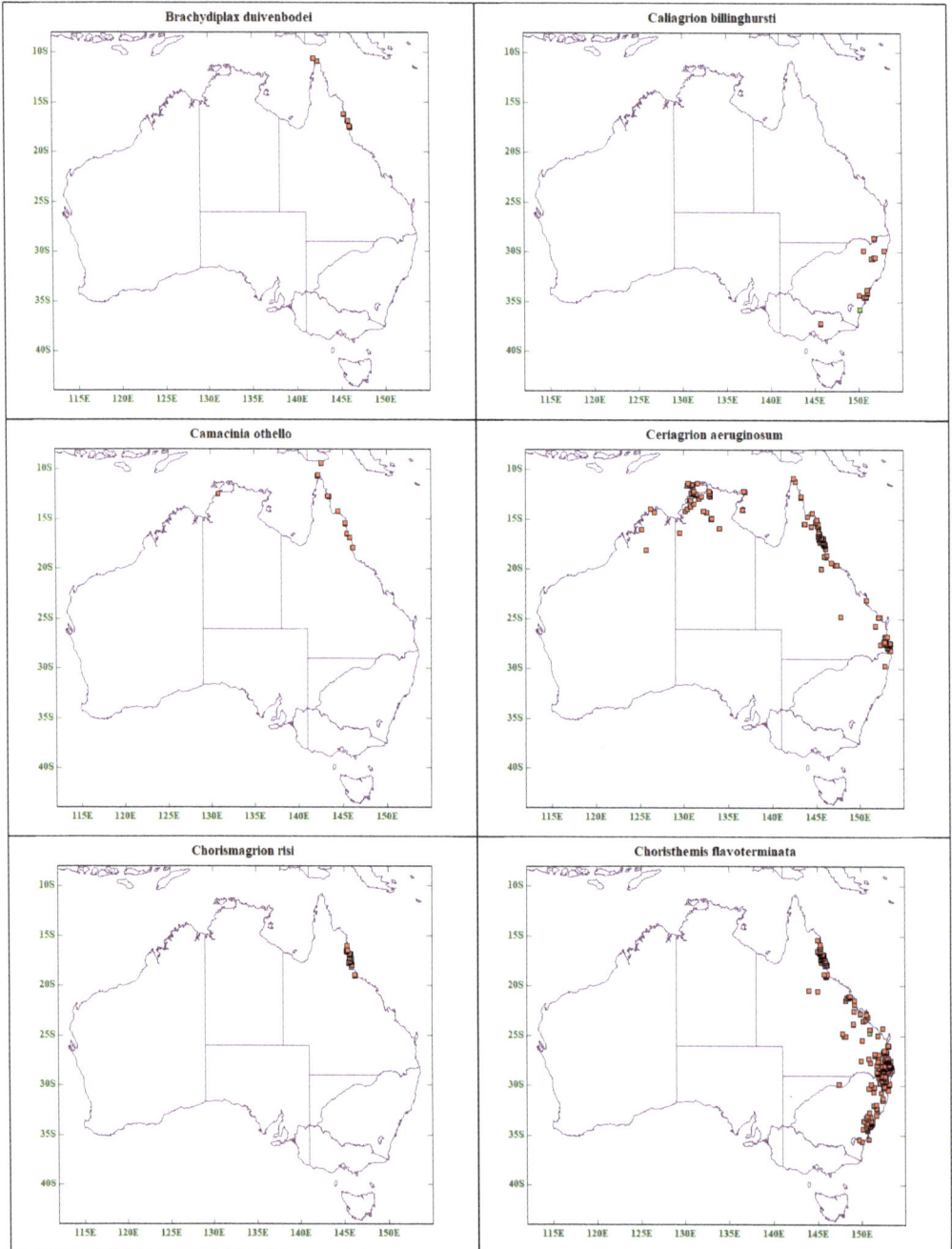

Brachydiplax duivenbodei

Caliagrion billinghursti

Camacinia othello

Ceriagrion aeruginosum

Chorismagrion risi

Choristhemis flavoterminata

Choristhemis olivei

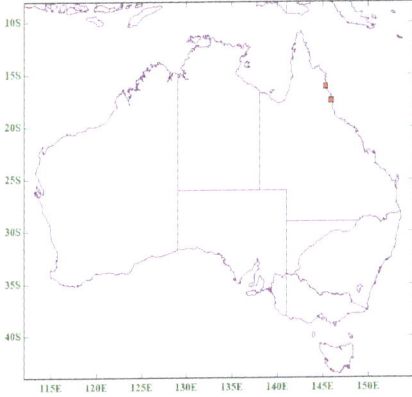

Cordulephya bidens

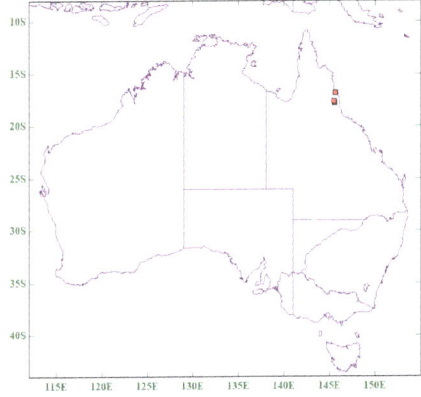

Cordulephya divergens

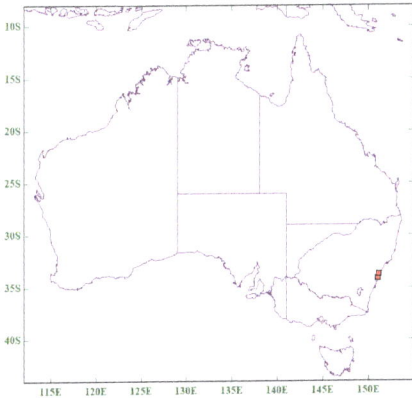

Cordulephya montana

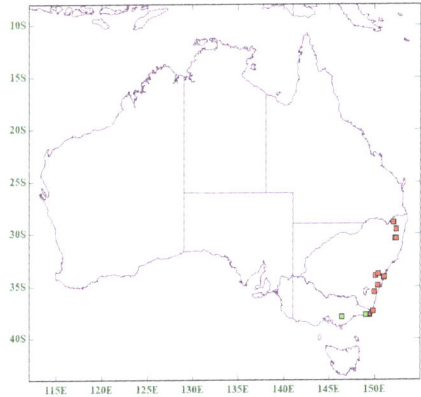

Cordulephya pygmaea

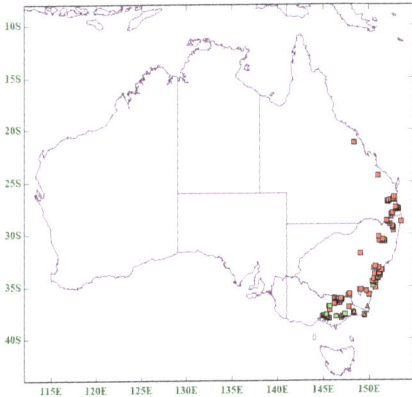

Crocothemis nigrifrons

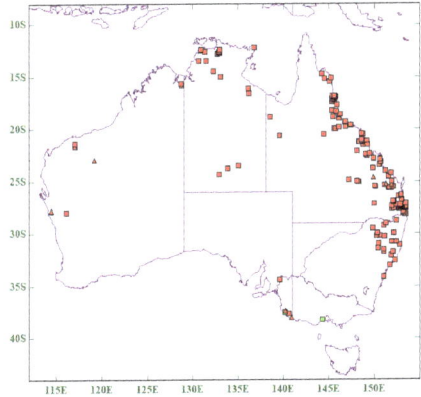

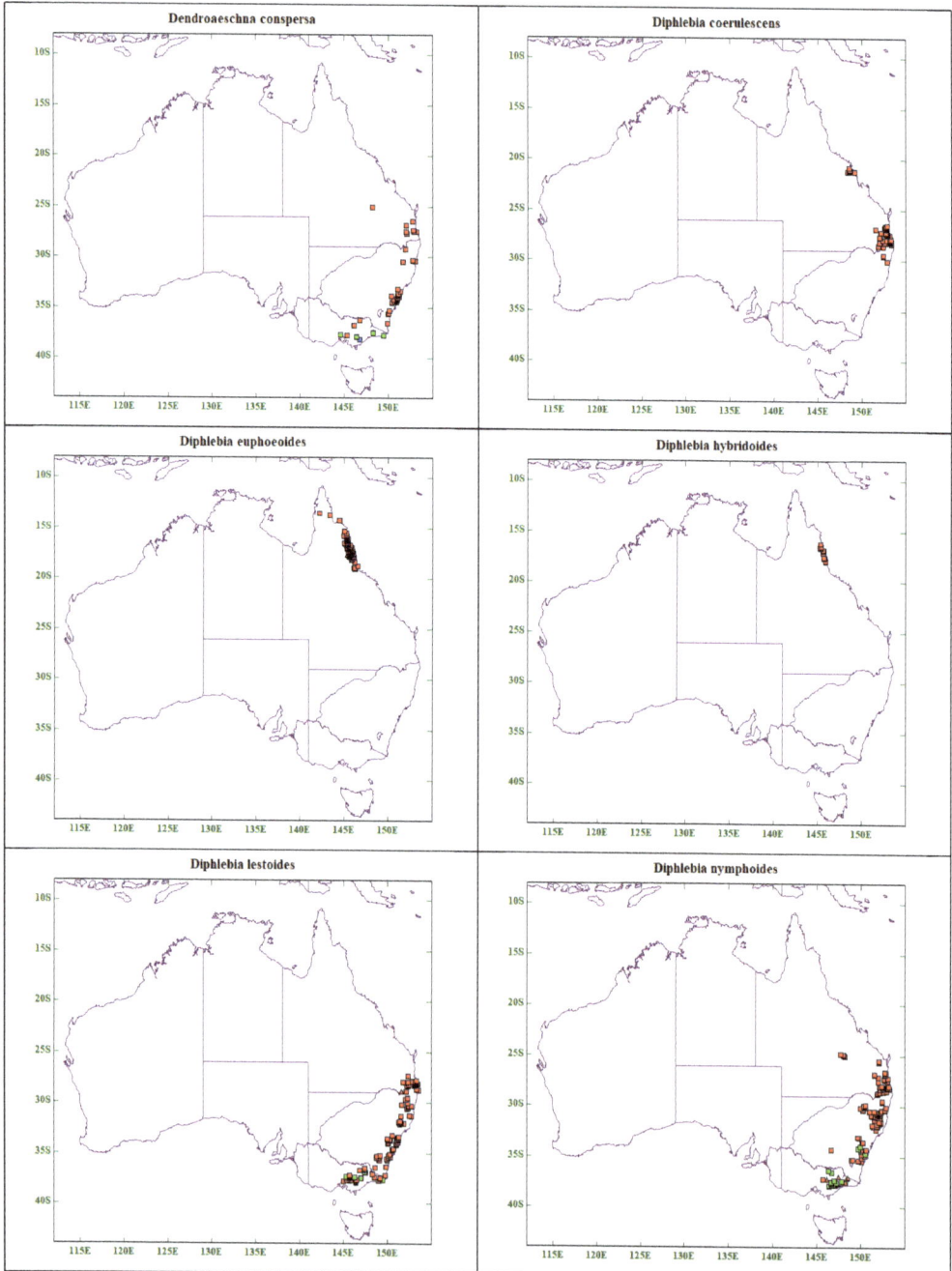

Dendroaeschna conspersa

Diphlebia coerulescens

Diphlebia euphoeoides

Diphlebia hybridoides

Diphlebia lestoides

Diphlebia nymphoides

Diplacodes bipunctata

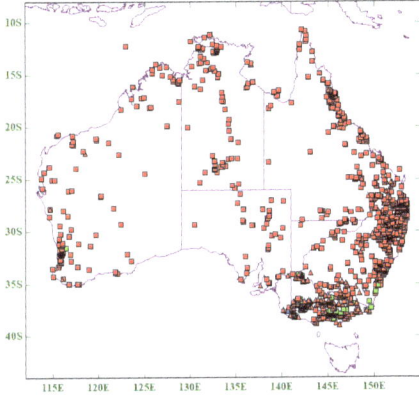

Diplacodes haematodes

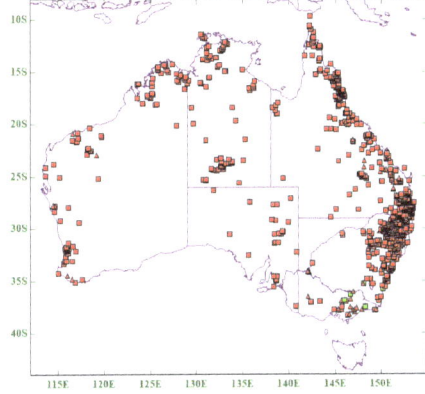

Diplacodes melanopsis

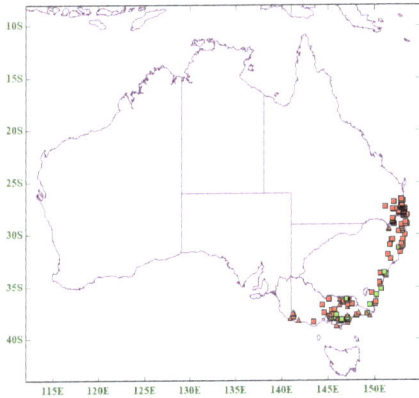

Diplacodes nebulosa

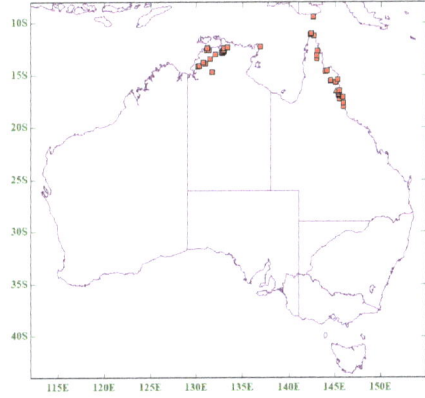

Diplacodes trivialis

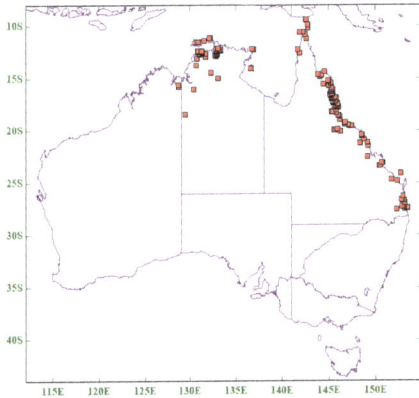

Dromaeschna weiskei

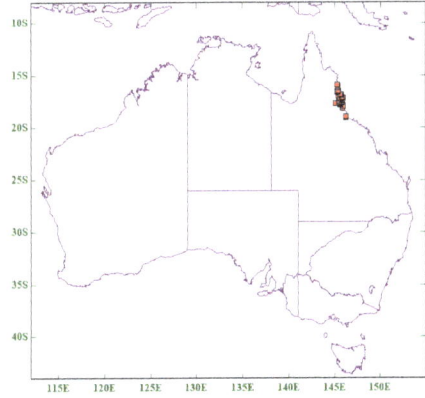

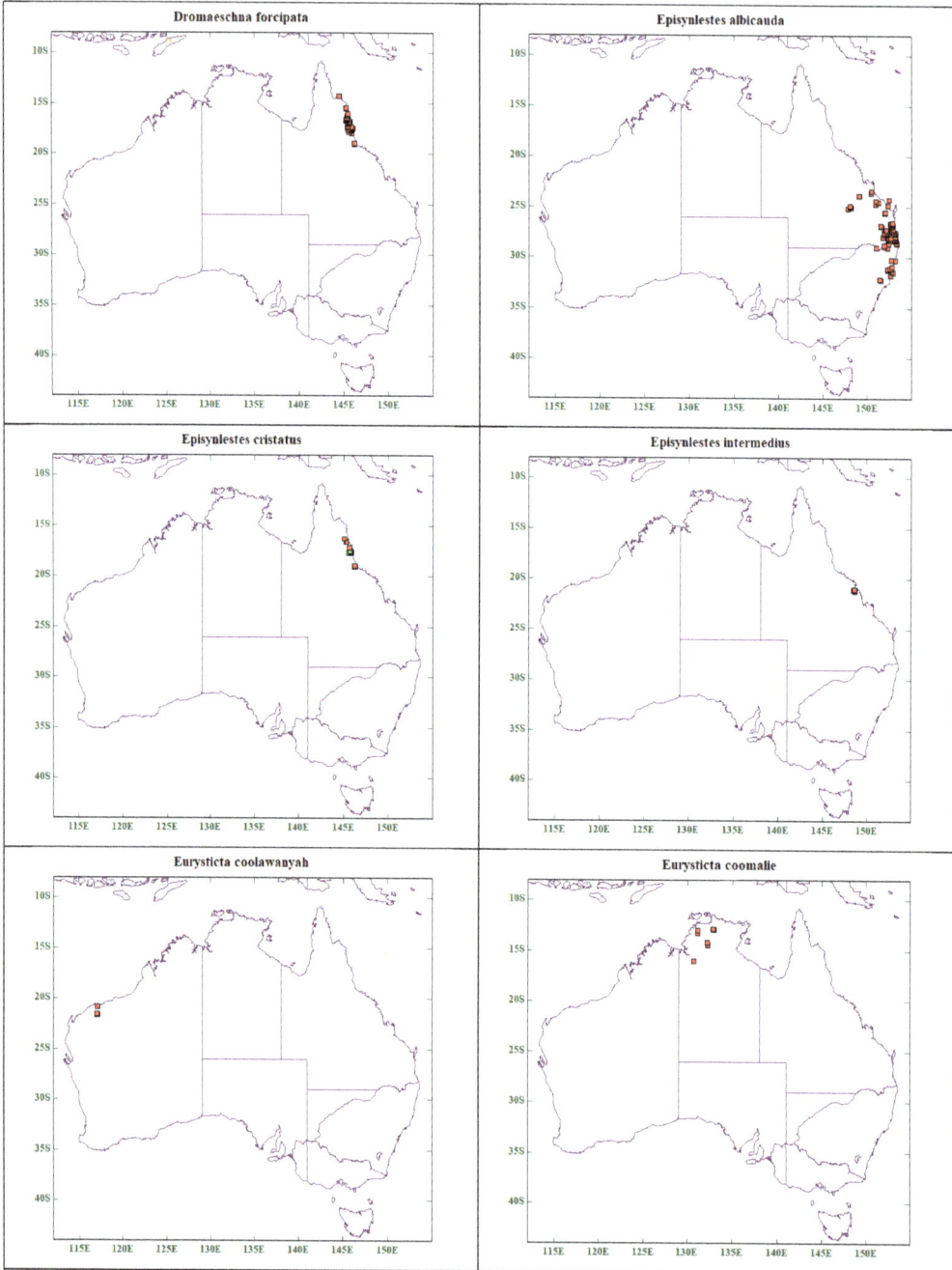

Dromaeschna forcipata

Episynlestes albicauda

Episynlestes cristatus

Episynlestes intermedius

Eurysticta coolawanyah

Eurysticta coomalie

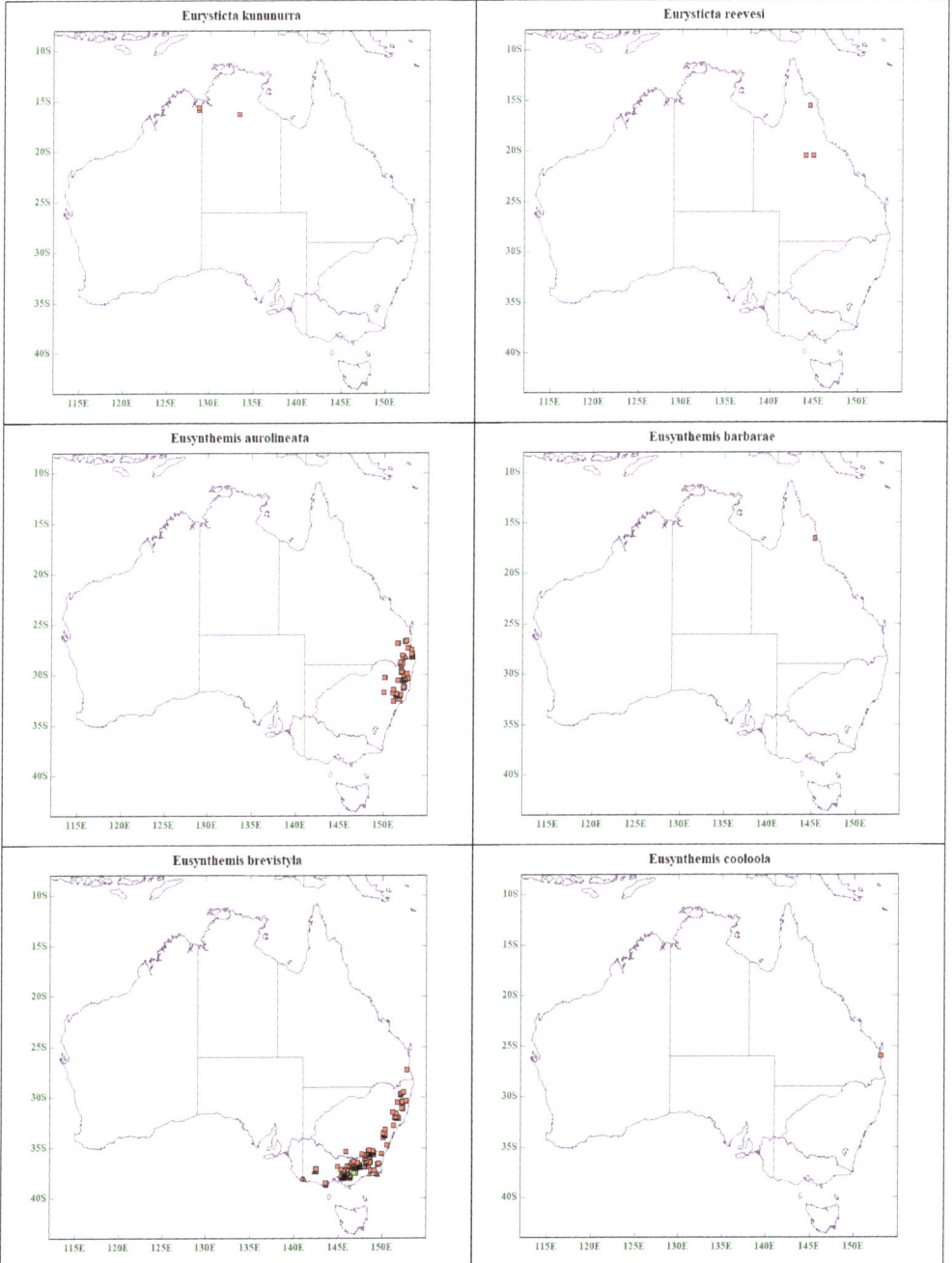

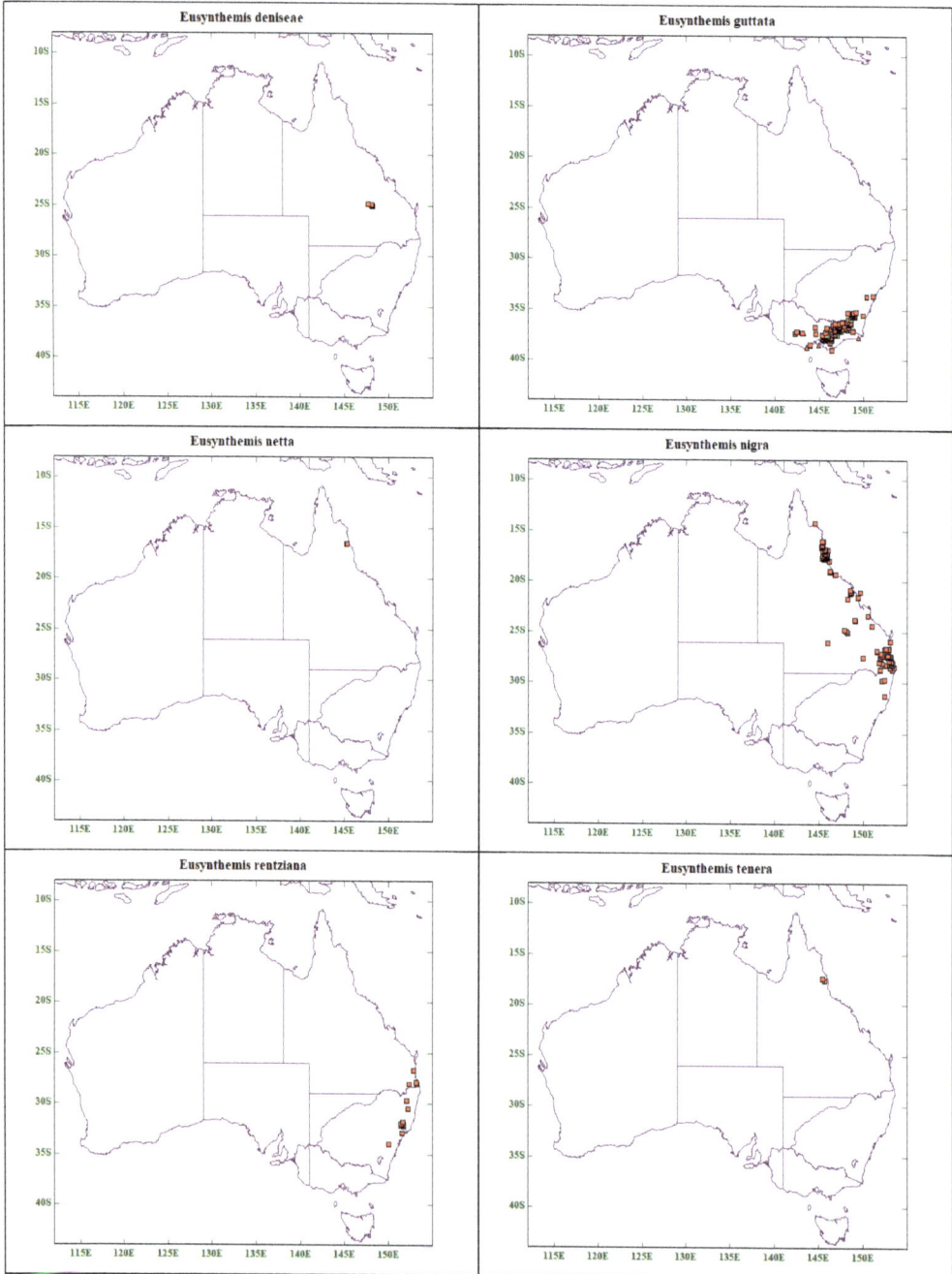

Eusynthemis deniseae

Eusynthemis guttata

Eusynthemis netta

Eusynthemis nigra

Eusynthemis rentziana

Eusynthemis tenera

Eusynthemis tillyardi

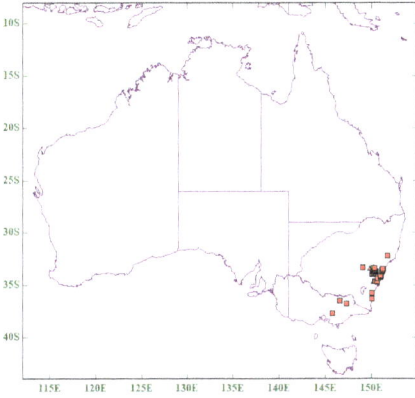

Eusynthemis ursa

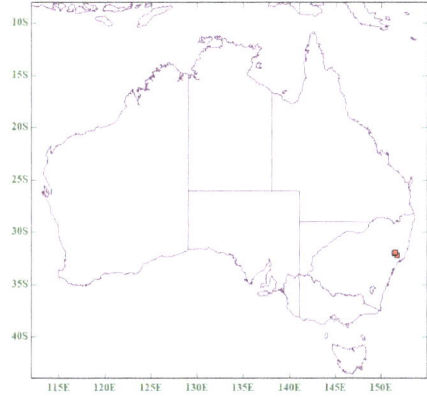

Eusynthemis ursula

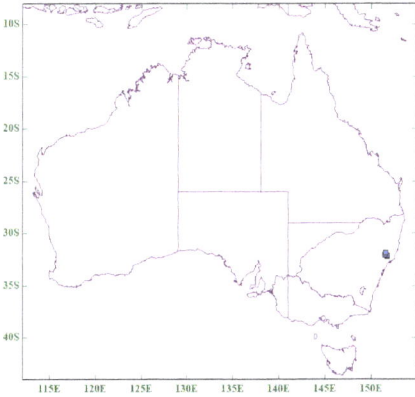

Eusynthemis virgula

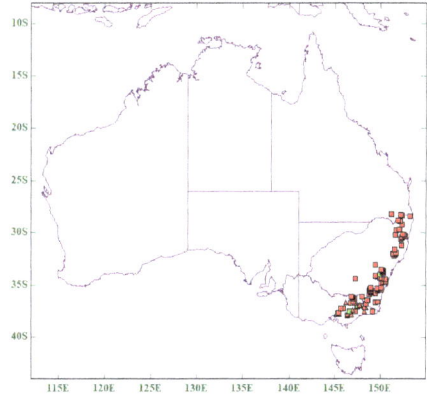

Griseargiolestes albescens

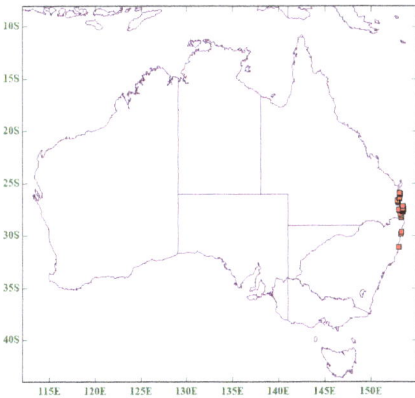

Griseargiolestes bucki

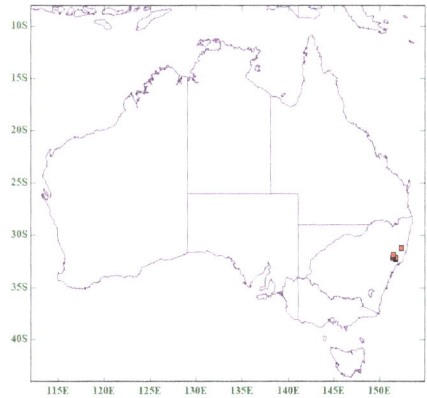

41

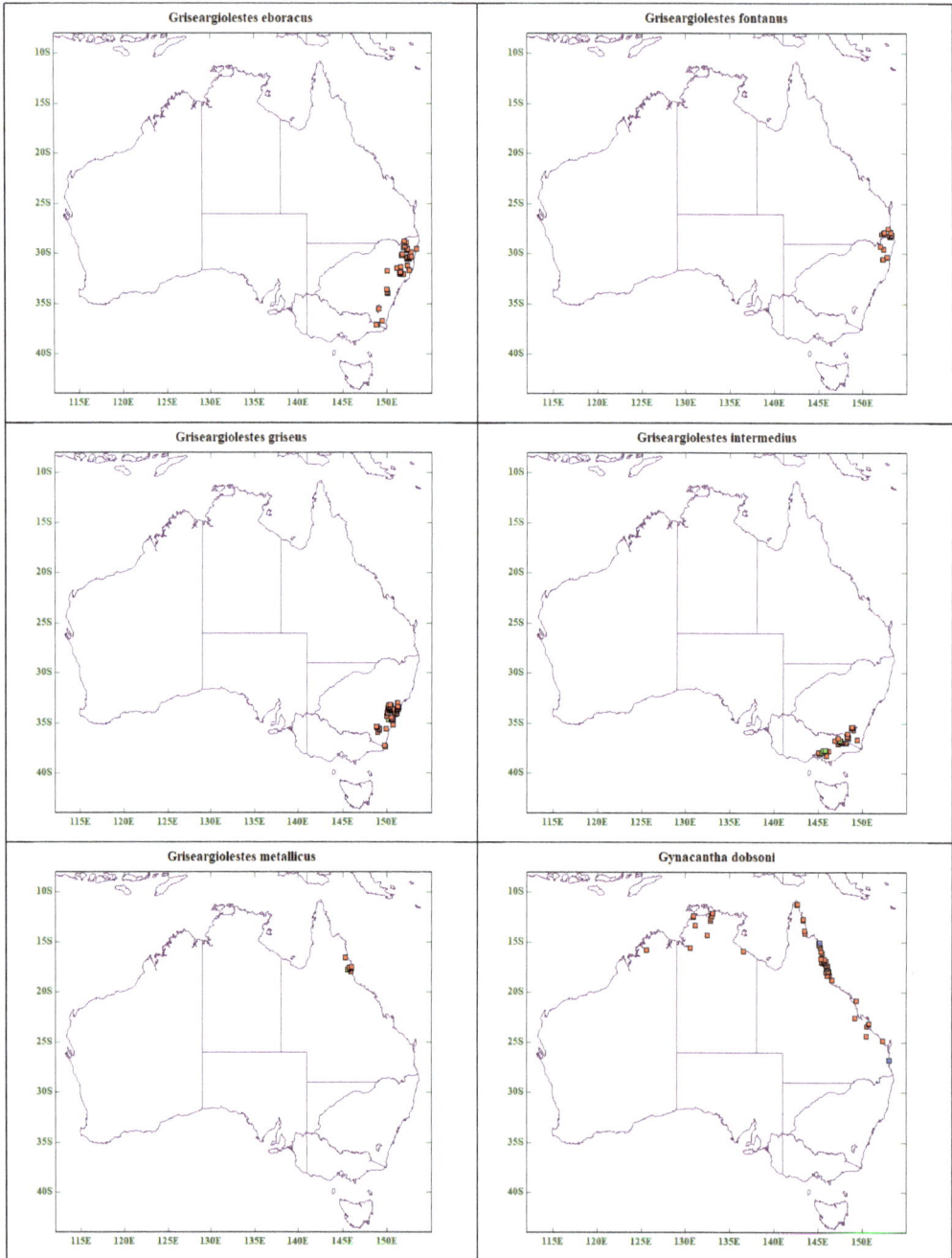

Griseargiolestes eboracus

Griseargiolestes fontanus

Griseargiolestes griseus

Griseargiolestes intermedius

Griseargiolestes metallicus

Gynacantha dobsoni

Gynacantha kirbyi

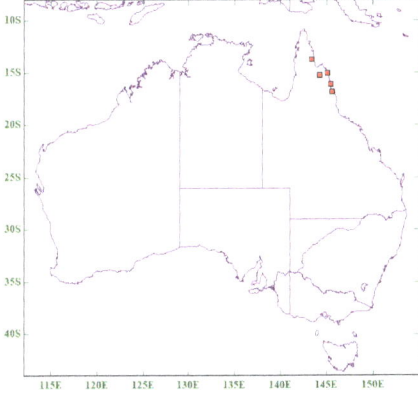

Gynacantha mocsaryi

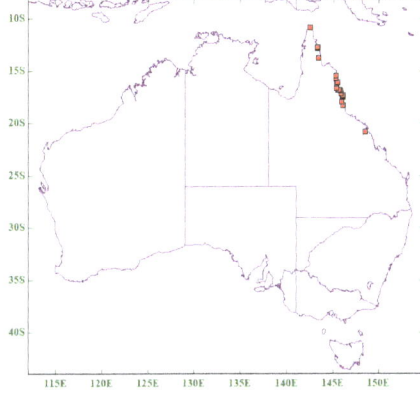

Gynacantha nourlangie

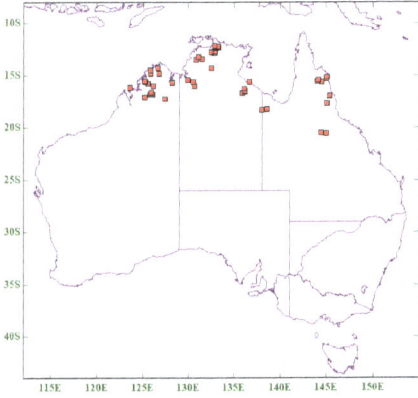

Gynacantha rosenbergi

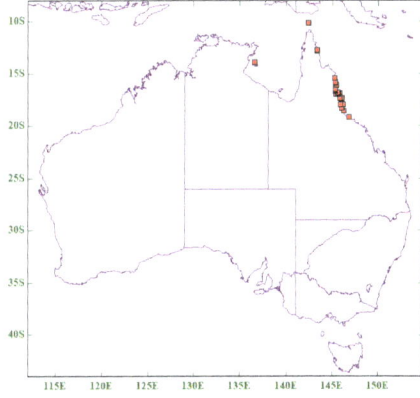

Hemianax papuensis

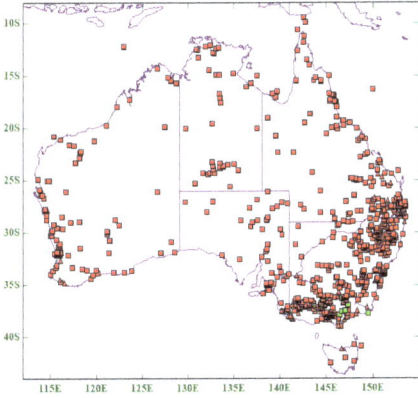

Hemicordulia australiae

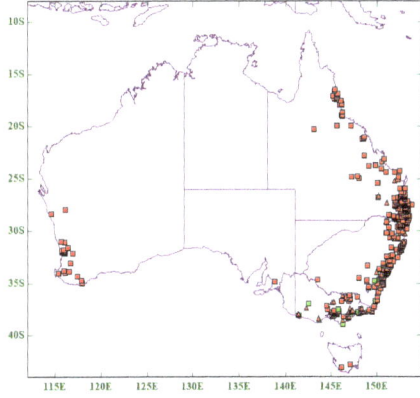

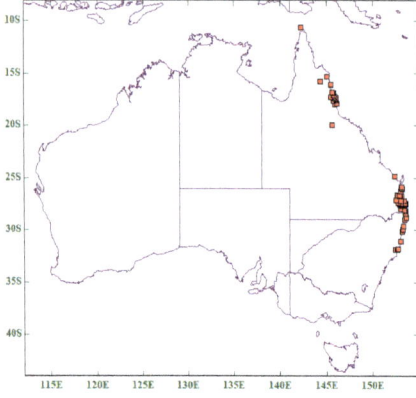

Hemicordulia continentalis

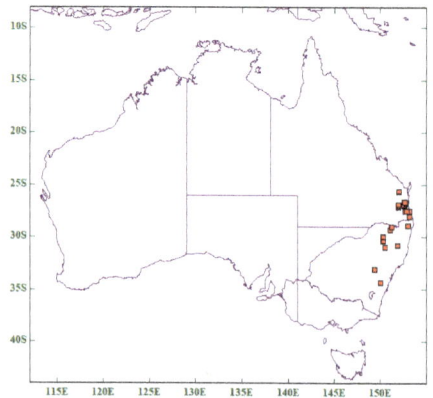

Hemicordulia flava

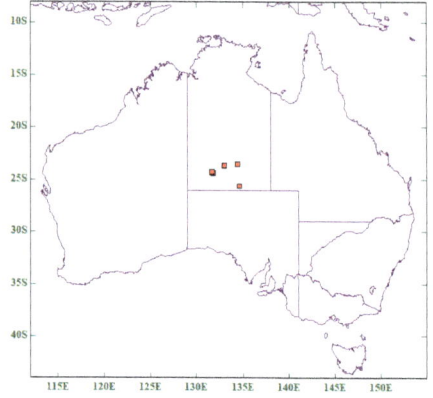

Hemicordulia intermedia

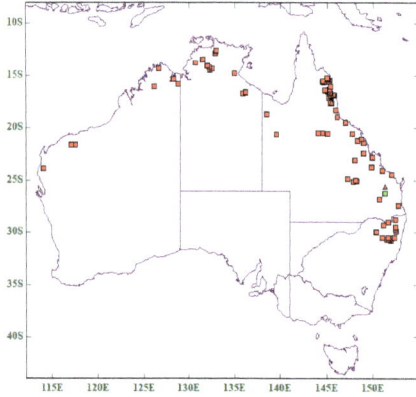

Hemicordulia kalliste

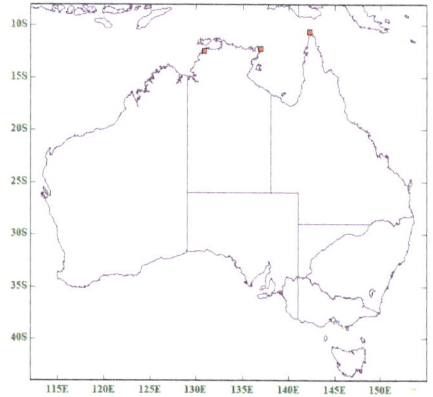

Hemicordulia koomina

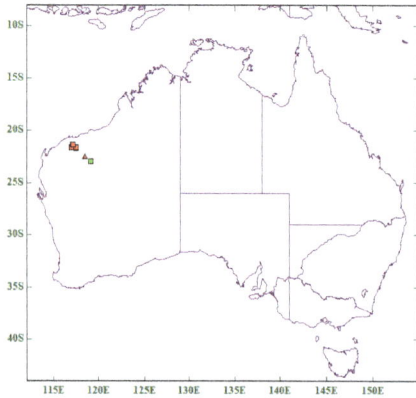

Hemicordulia superba

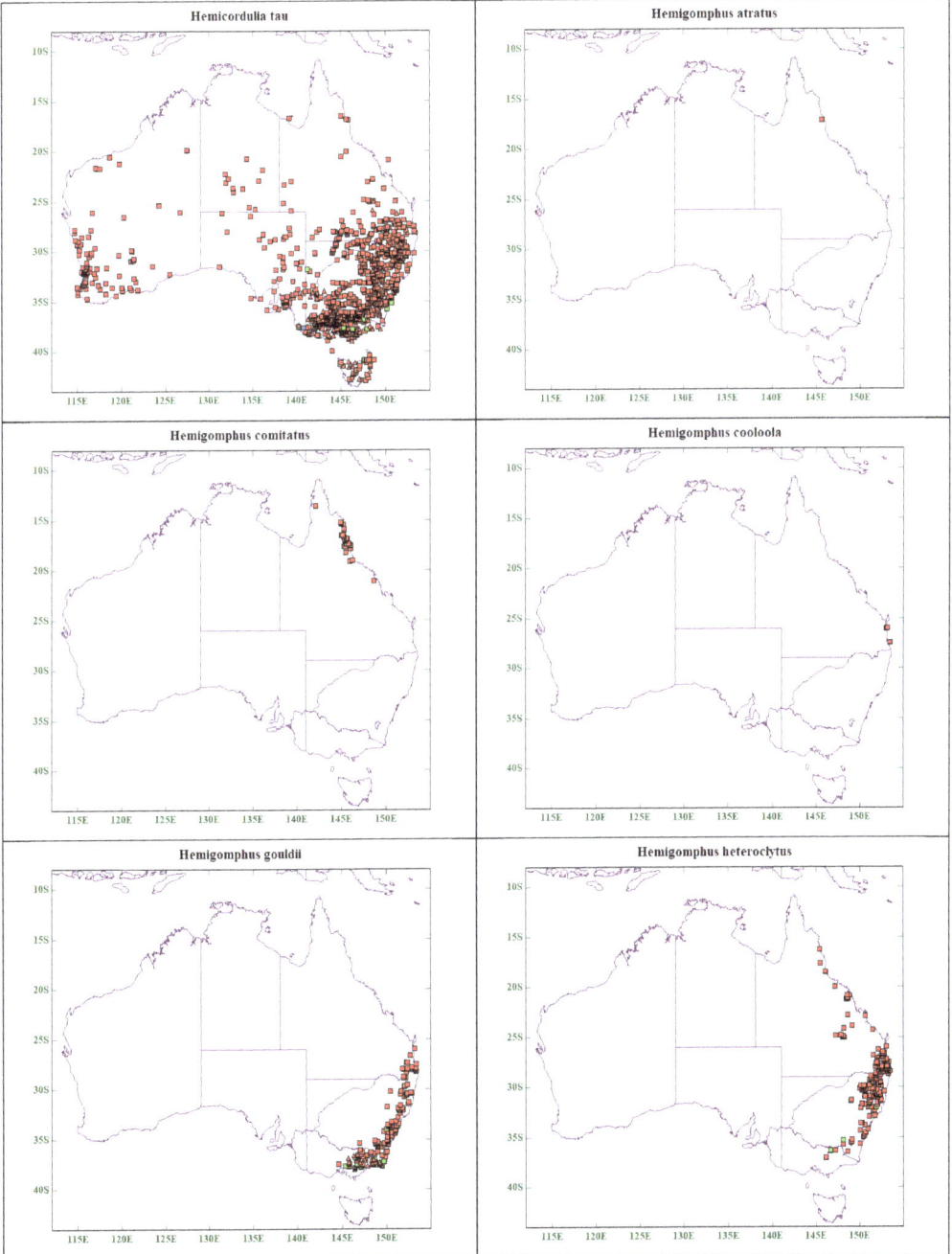

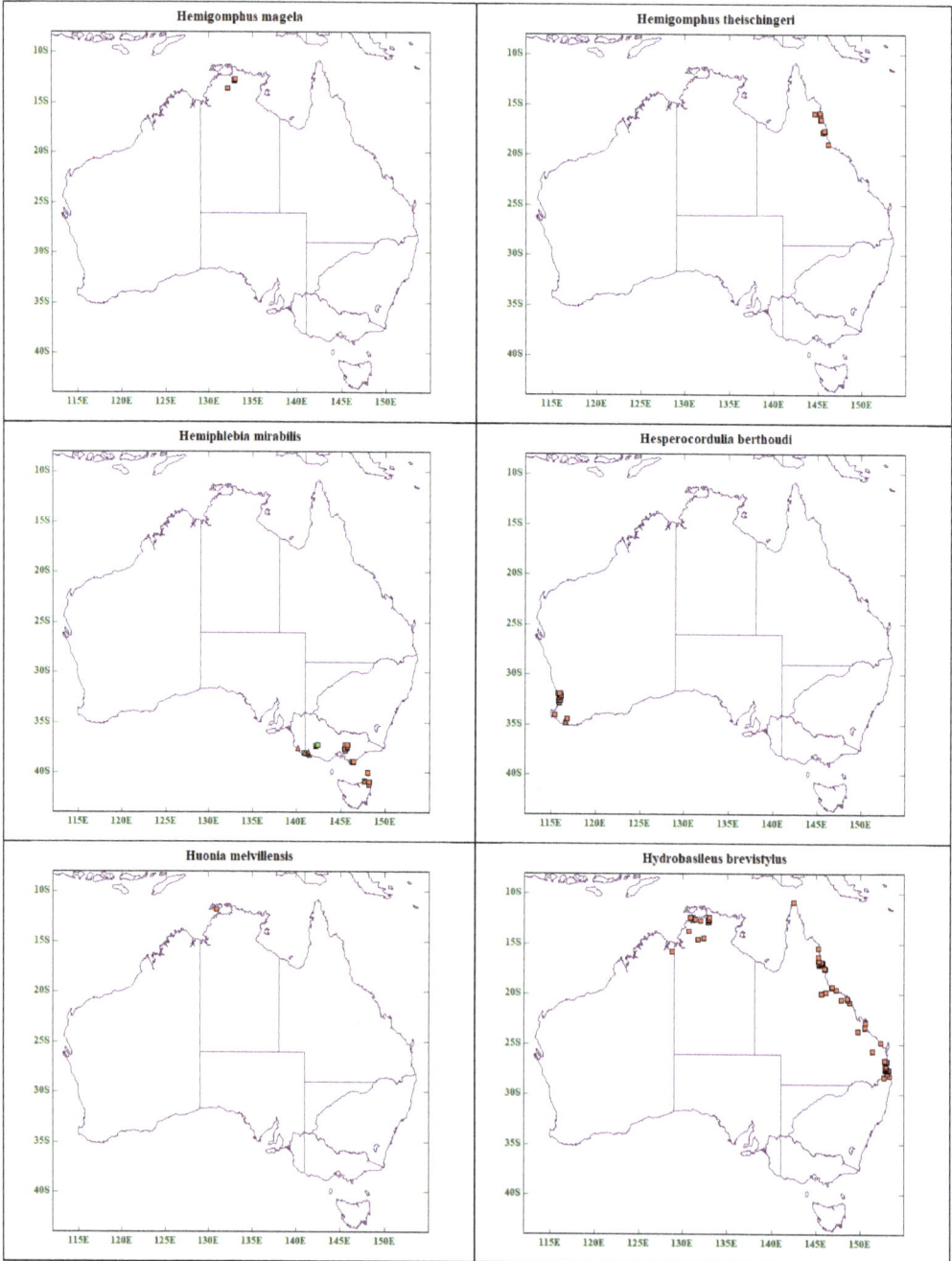

Hemigomphus magela

Hemigomphus theischingeri

Hemiphlebia mirabilis

Hesperocordulia berthoudi

Huonia melvillensis

Hydrobasileus brevistylus

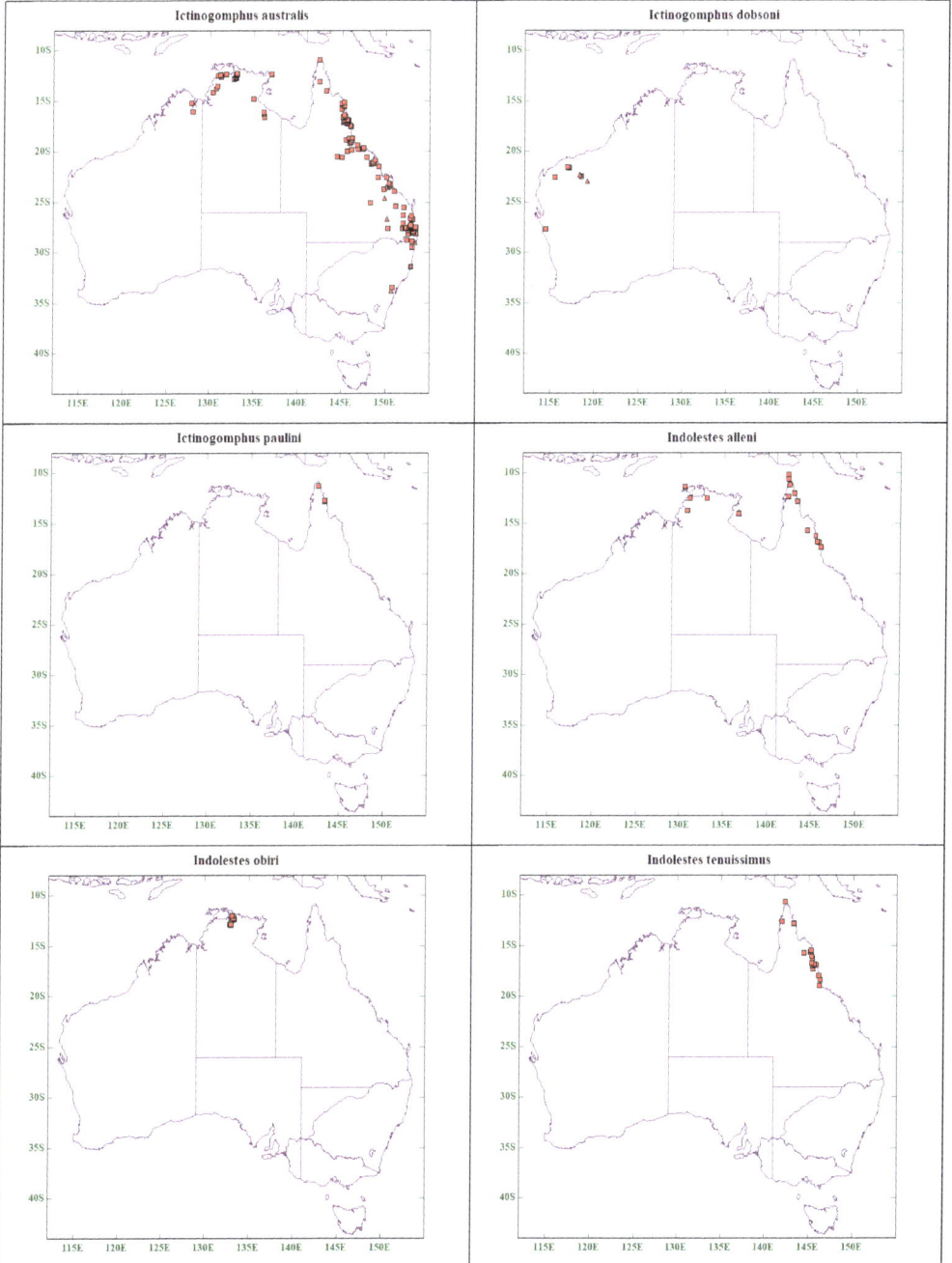

47

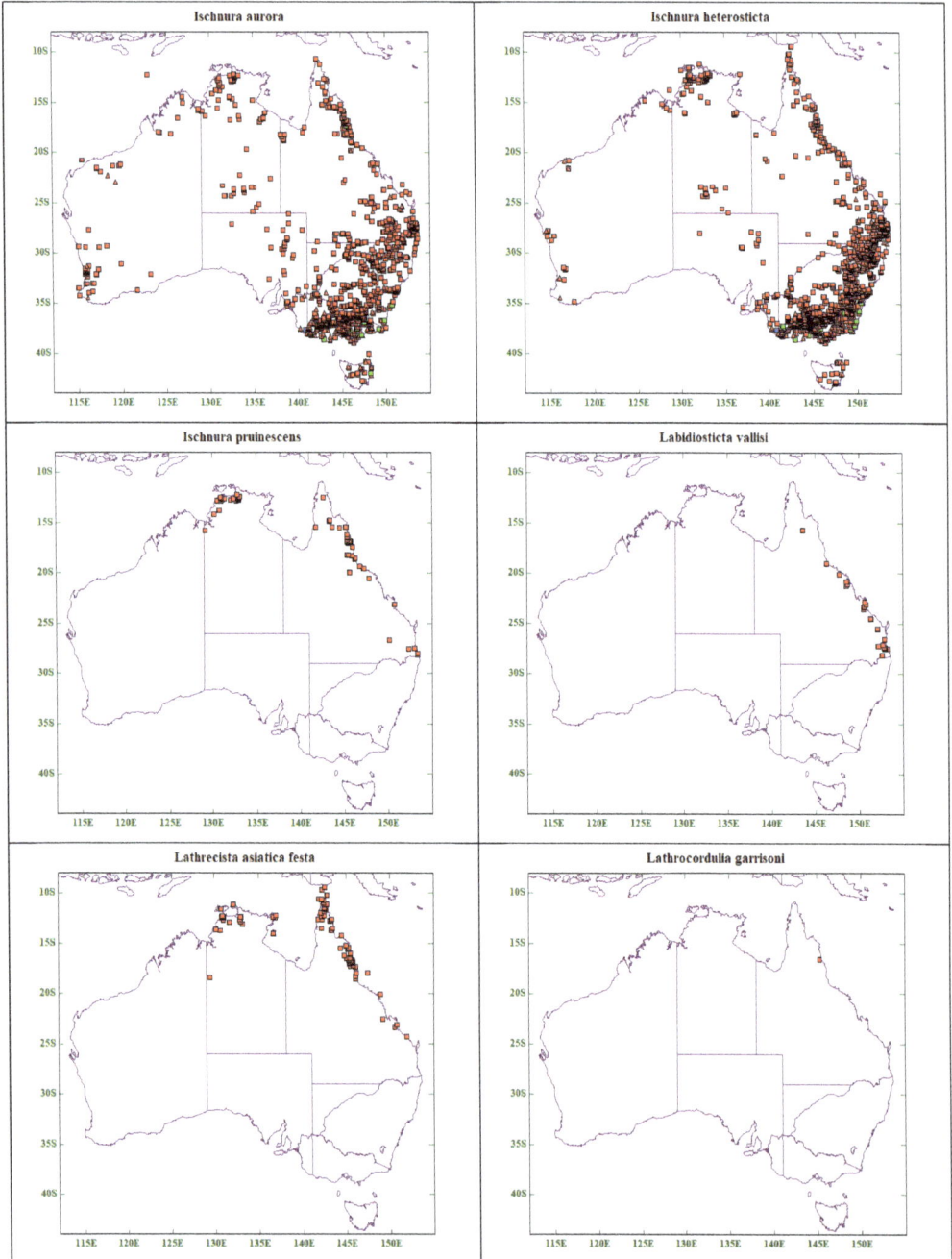

Ischnura aurora

Ischnura heterosticta

Ischnura pruinescens

Labidiosticta vallisi

Lathrecista asiatica festa

Lathrocordulia garrisoni

Lathrocordulia metallica

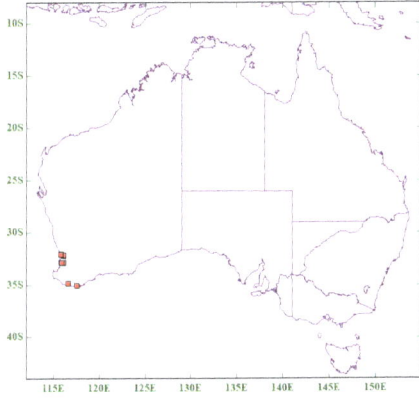

Lestes concinnus

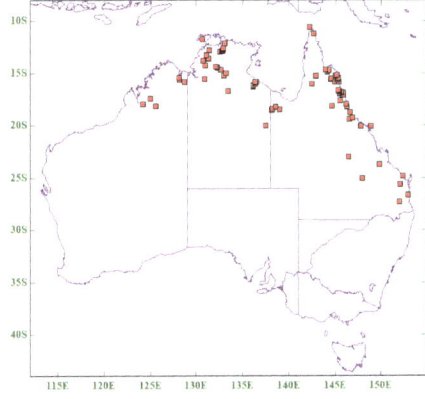

Lestoidea barbarae

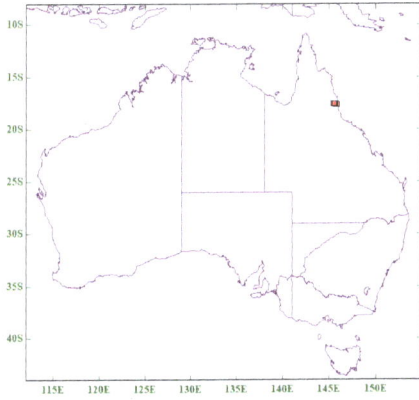

Lestoidea brevicauda

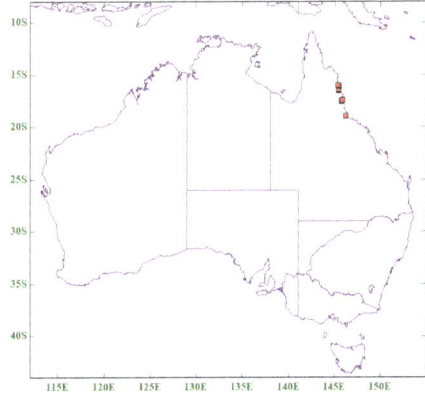

Lestoidea conjuncta

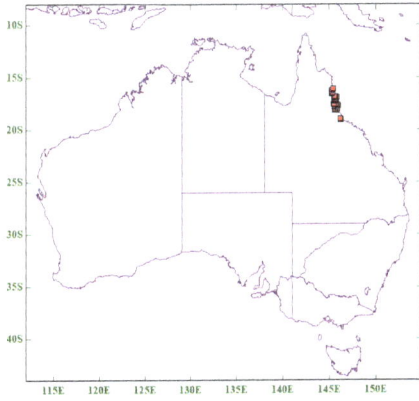

Lestoidea lewisiana

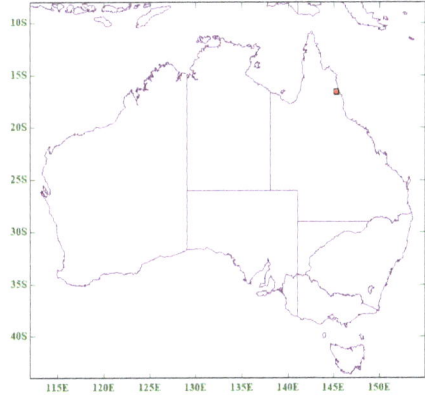

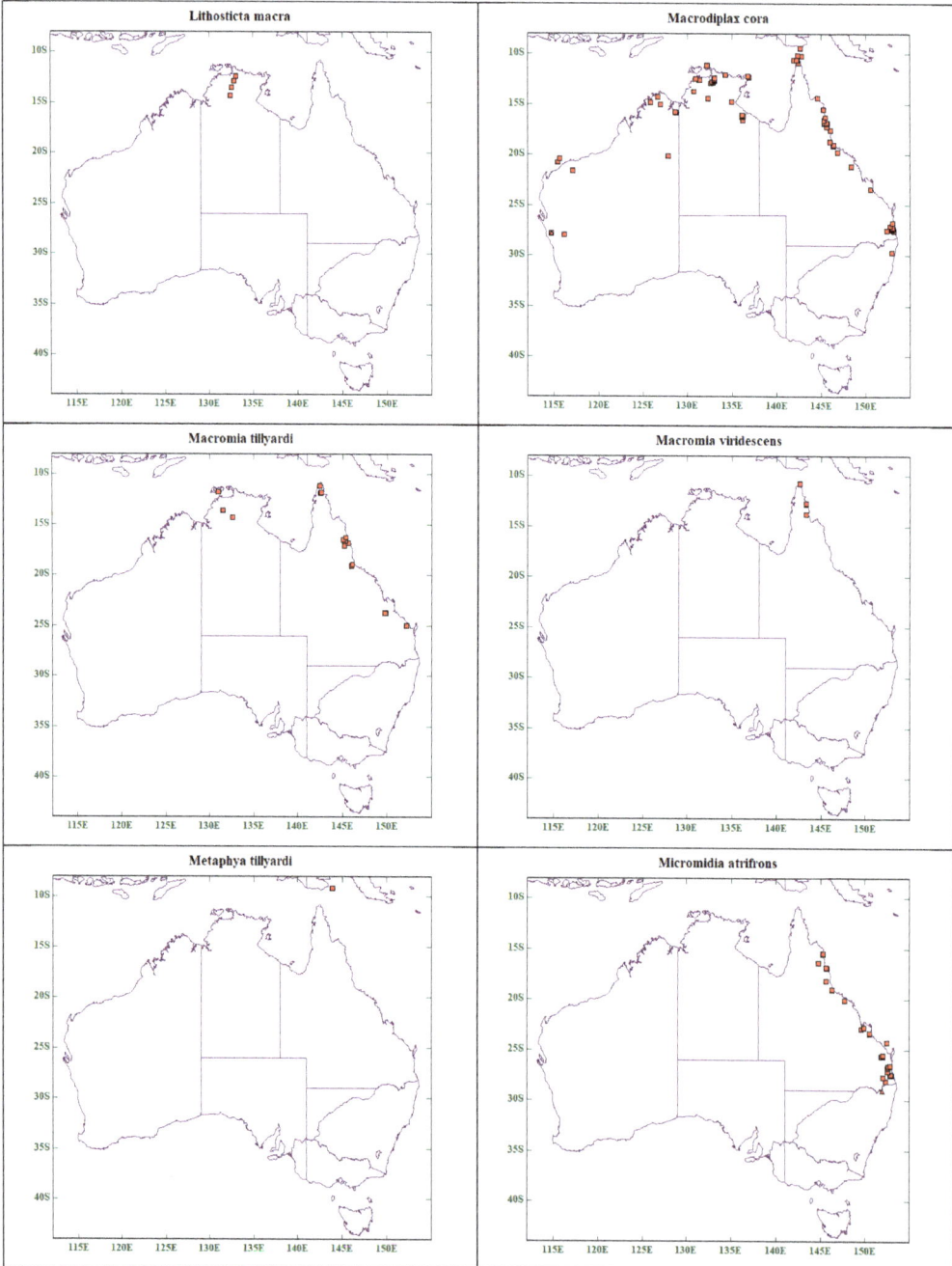

Lithosticta macra

Macrodiplax cora

Macromia tillyardi

Macromia viridescens

Metaphya tillyardi

Micromidia atrifrons

Micromidia convergens

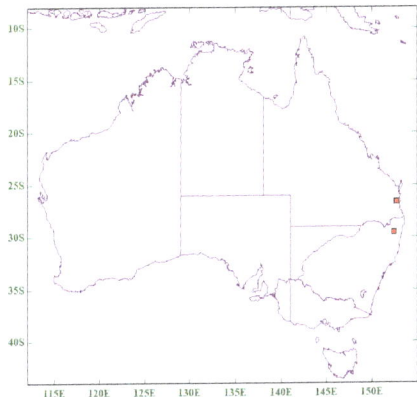

Micromidia rodericki

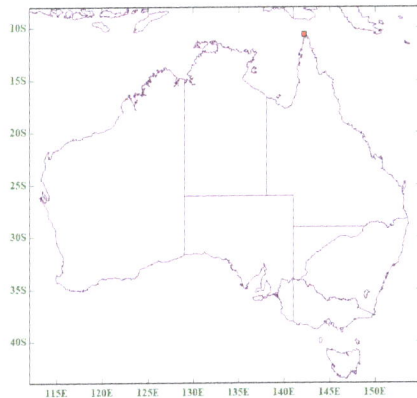

Miniargiolestes minimus

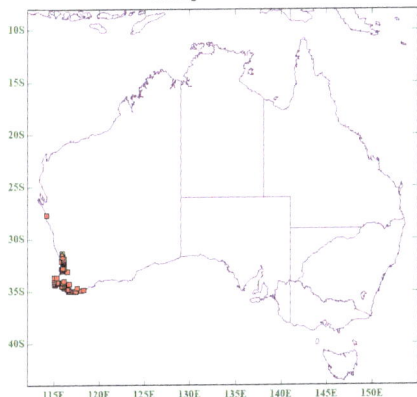

Nannodiplax rubra

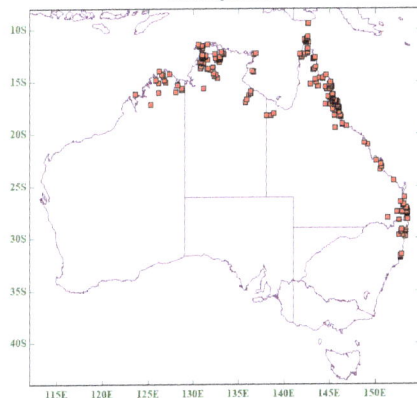

Nannophlebia eludens

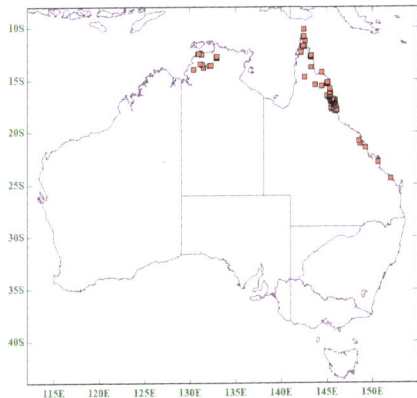

Nannophlebia injibandi

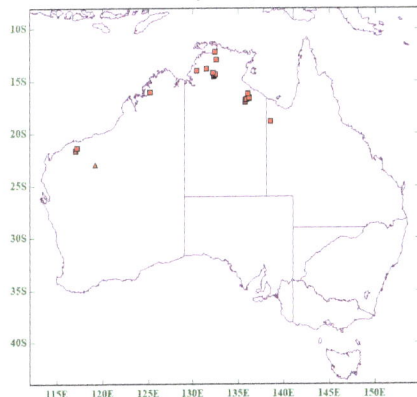

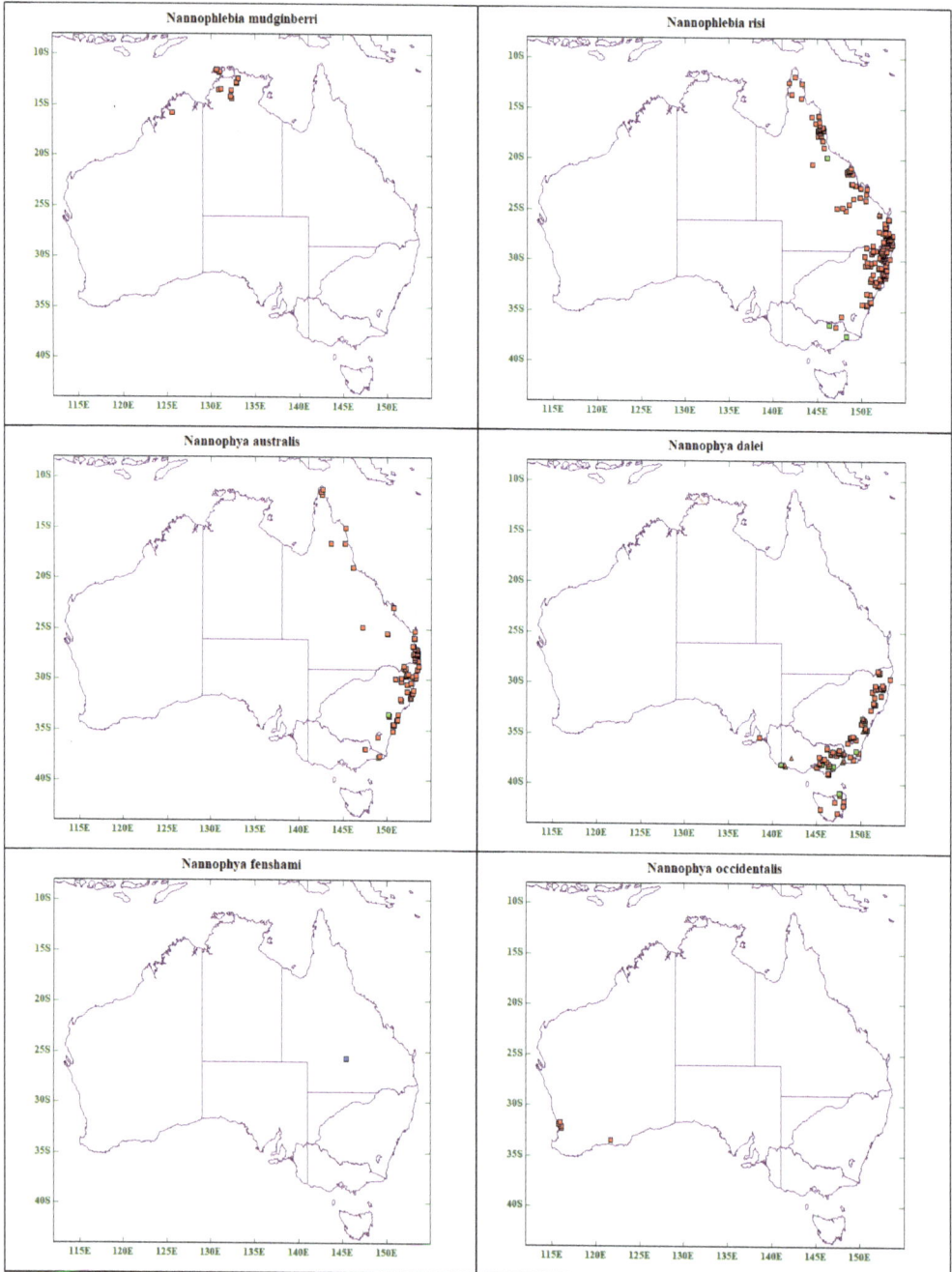

Nannophlebia mudginberri

Nannophlebia risi

Nannophya australis

Nannophya dalei

Nannophya fenshami

Nannophya occidentalis

Nannophya paulsoni

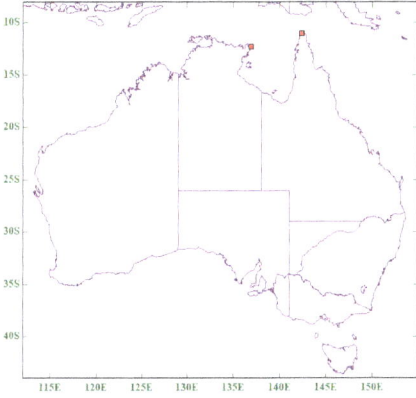

Neosticta canescens

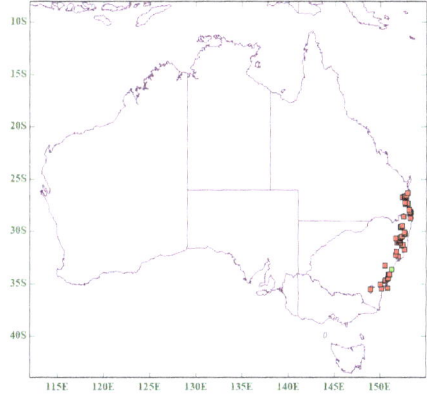

Neosticta fraseri

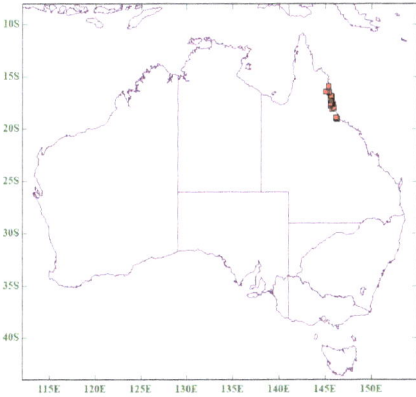

Neosticta silvarum

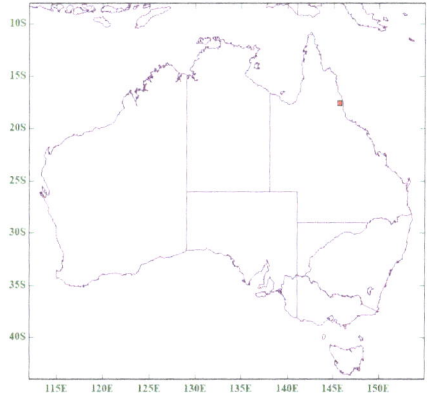

Neurothemis oligoneura

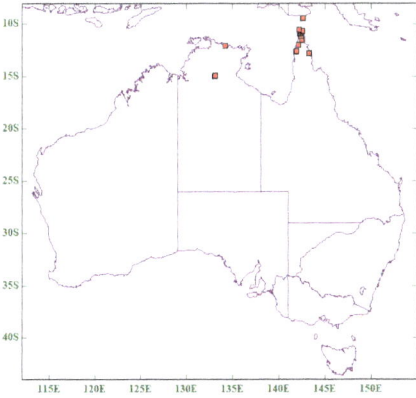

Neurothemis stigmatizans

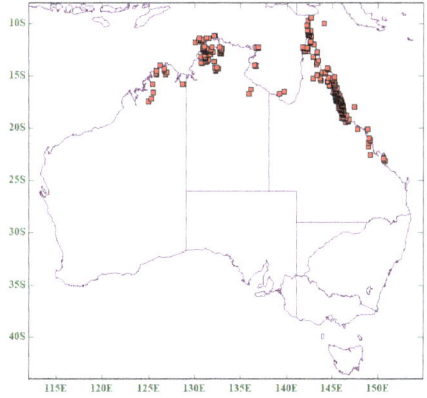

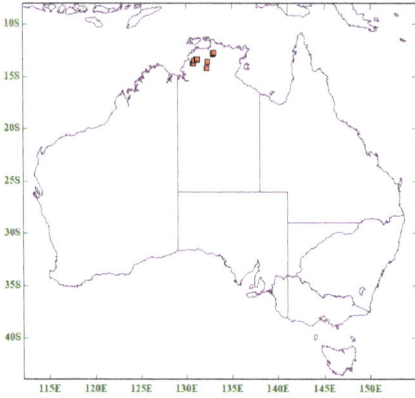

Nososticta baroalba

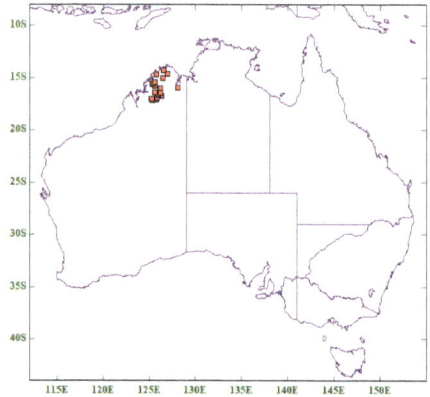

Nososticta coelestina

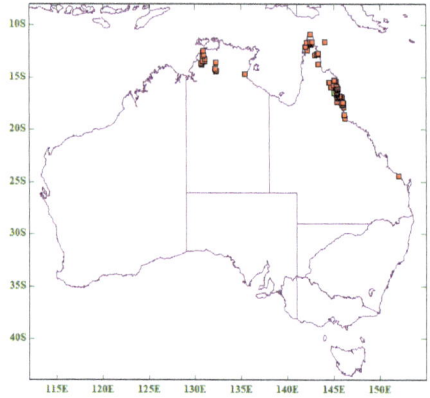

Nososticta fraterna

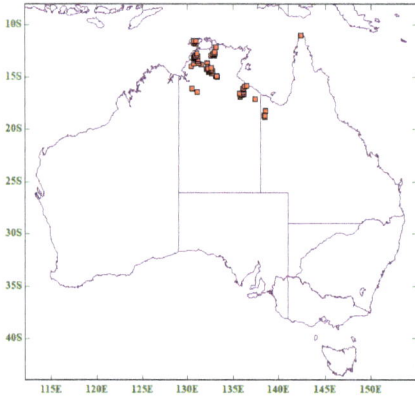

Nososticta kalumburu

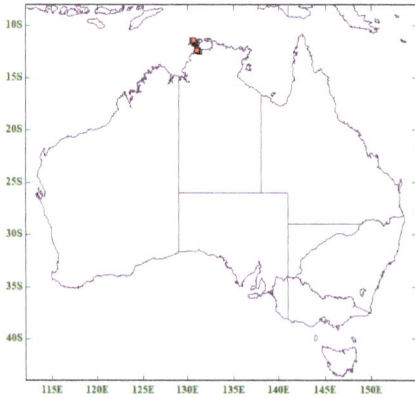

Nososticta koolpinyah

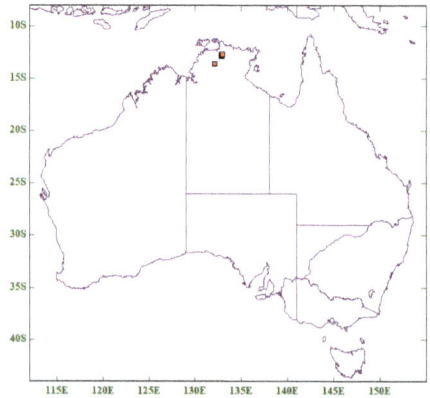

Nososticta koongarra

Nososticta liveringa

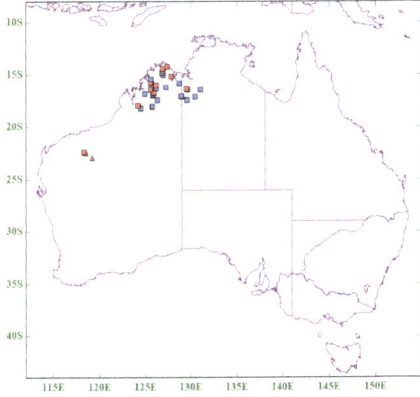

Nososticta mouldsi

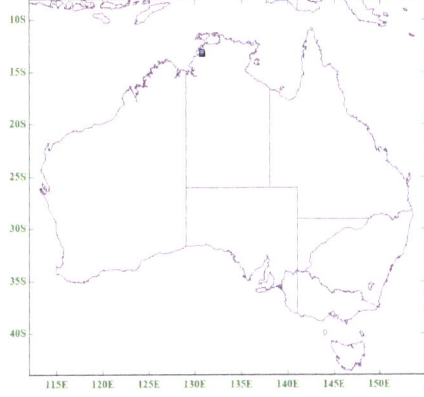

Nososticta pilbara

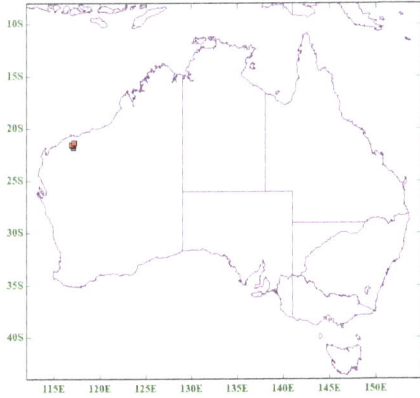

Nososticta solida

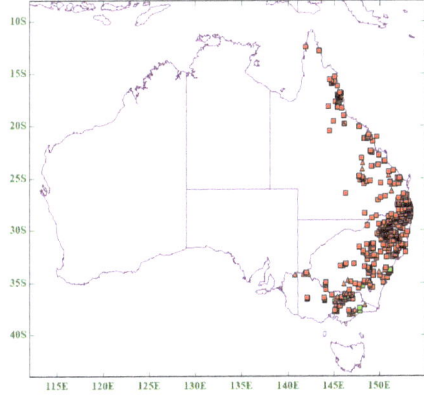

Nososticta solitaria

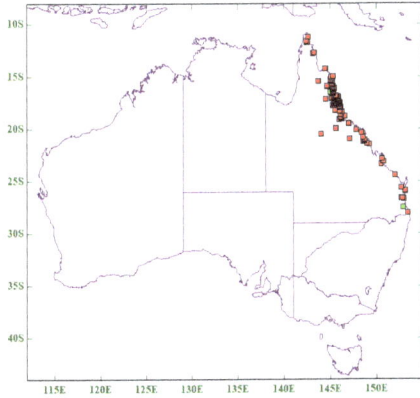

Nososticta taracumbi

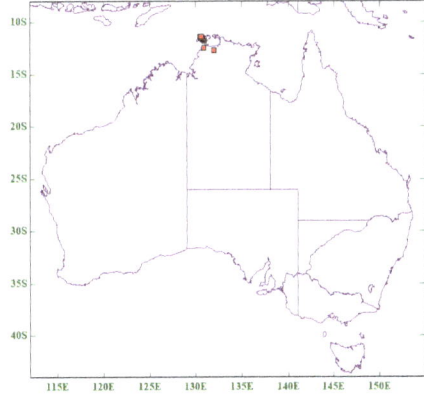

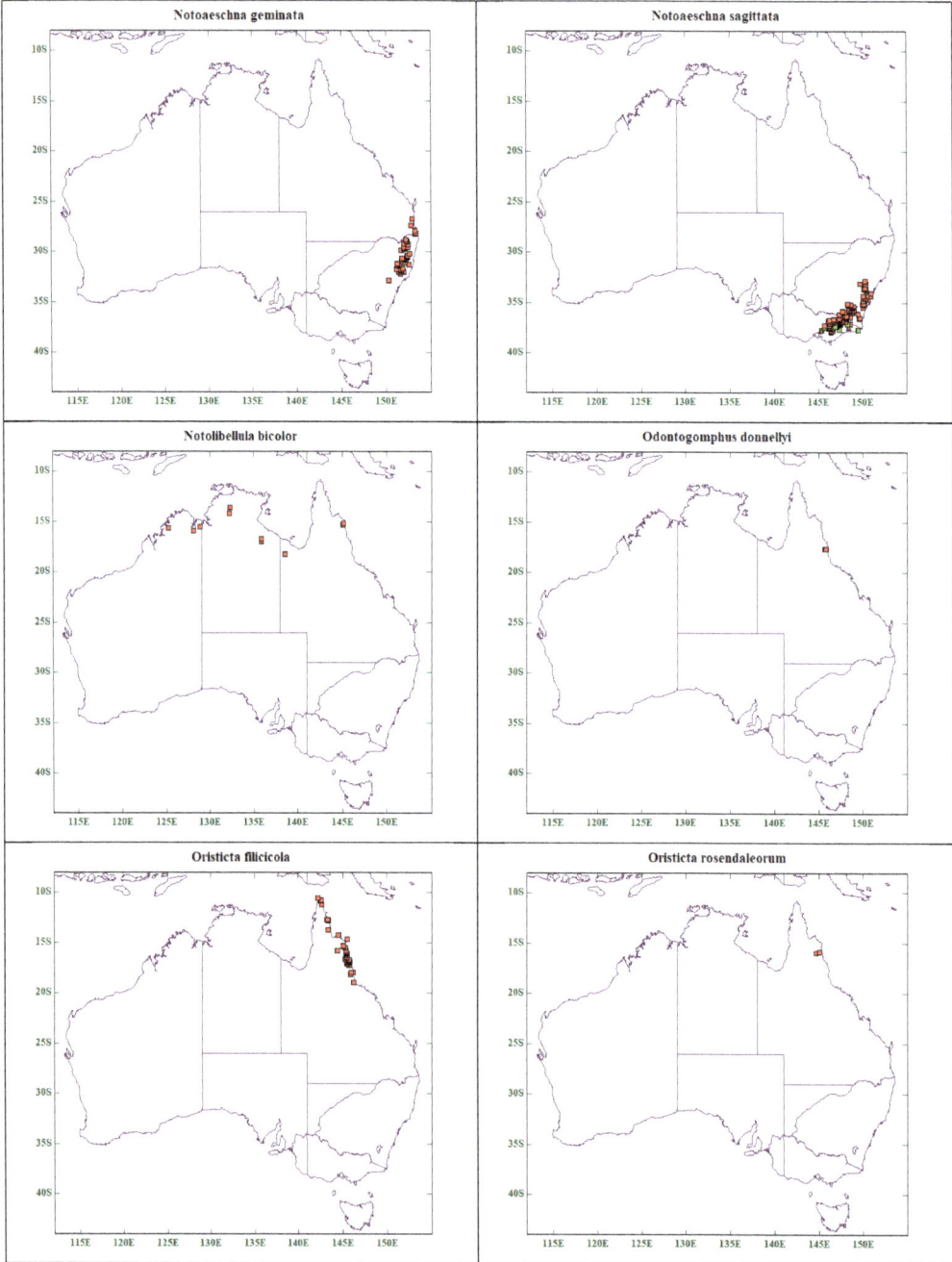

Notoaeschna geminata

Notoaeschna sagittata

Notolibellula bicolor

Odontogomphus donnellyi

Oristicta filicicola

Oristicta rosendaleorum

Orthetrum balteatum

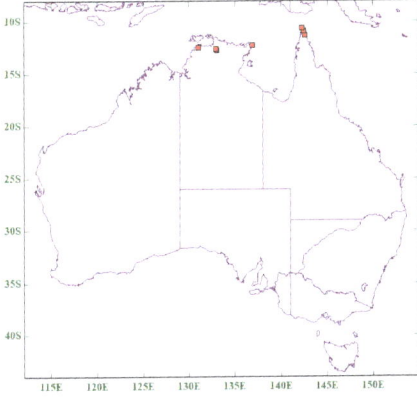

Orthetrum boumiera

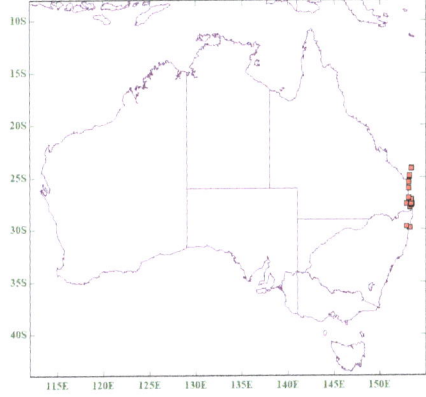

Orthetrum caledonicum

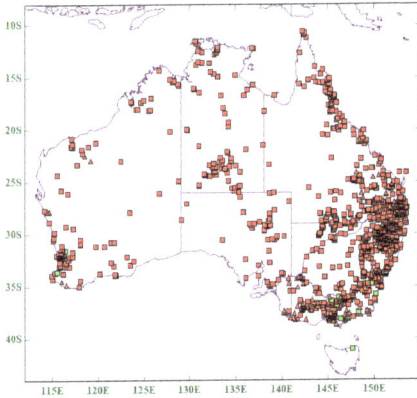

Orthetrum migratum

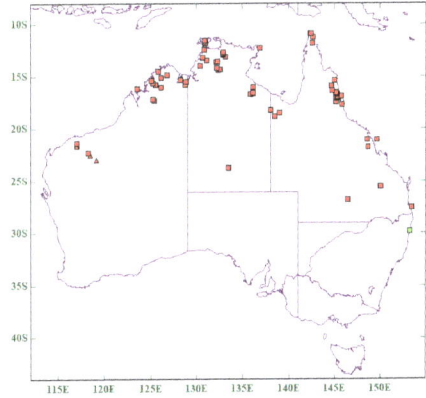

Orthetrum sabina

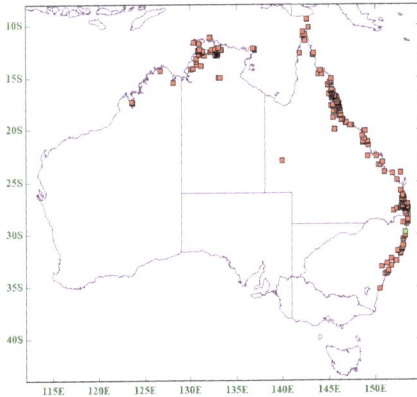

Orthetrum serapia

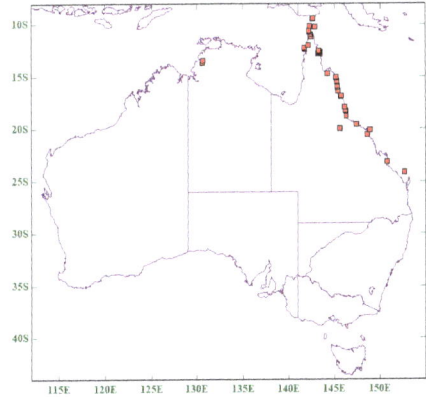

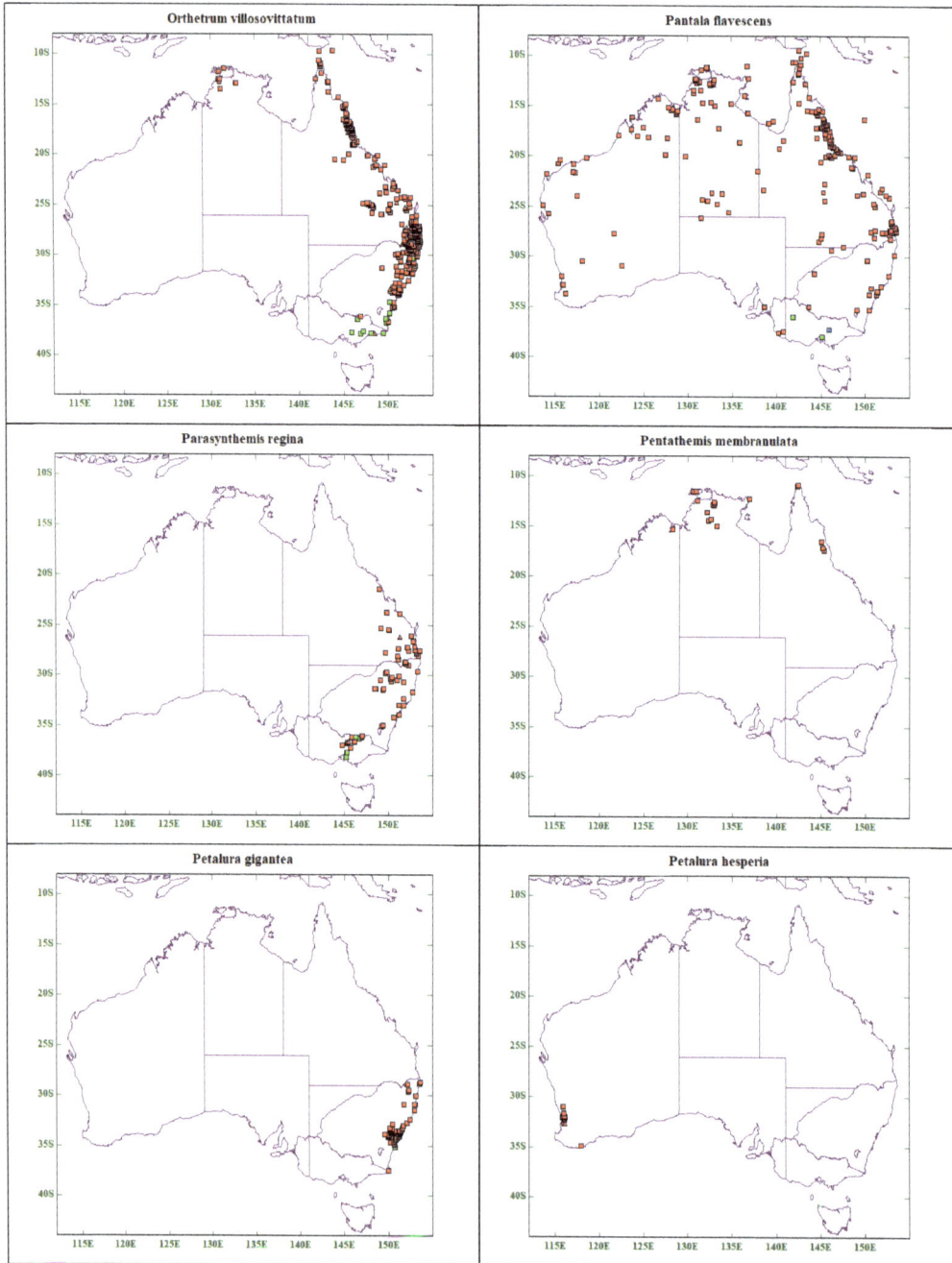

Orthetrum villosovittatum

Pantala flavescens

Parasynthemis regina

Pentathemis membranulata

Petalura gigantea

Petalura hesperia

Petalura ingentissima

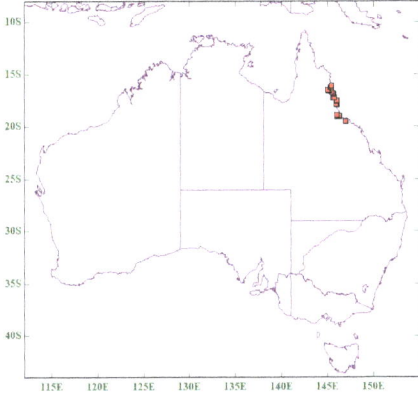

Petalura litorea

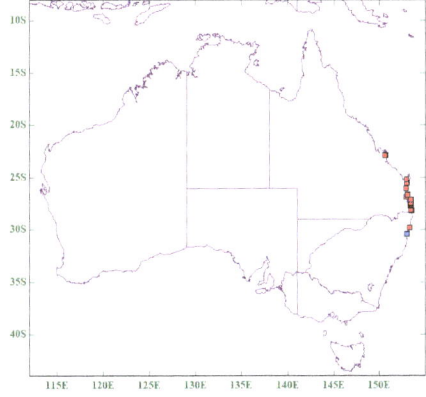

Petalura pulcherrima

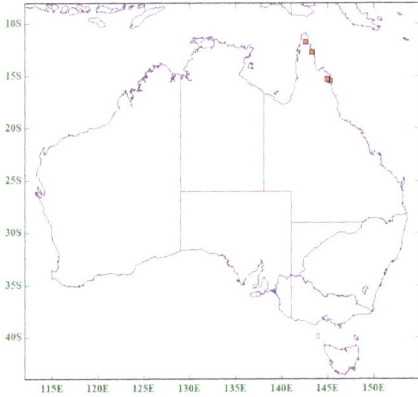

Podopteryx selysi

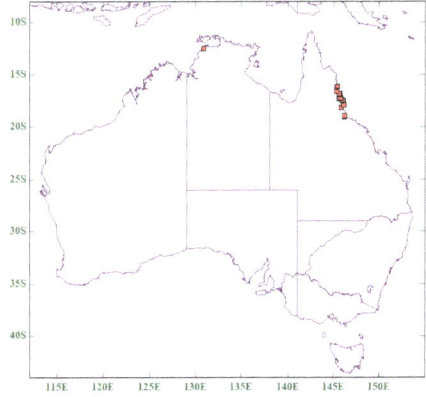

Potamarcha congener

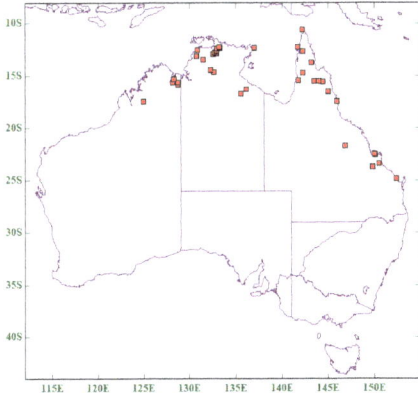

Procordulia affinis

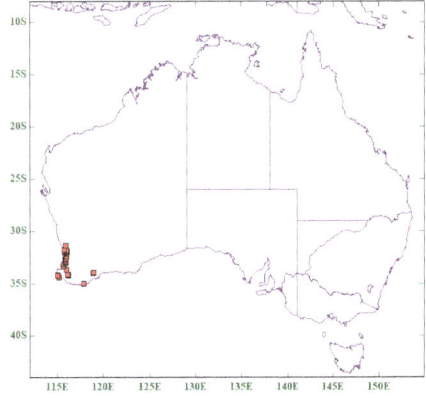

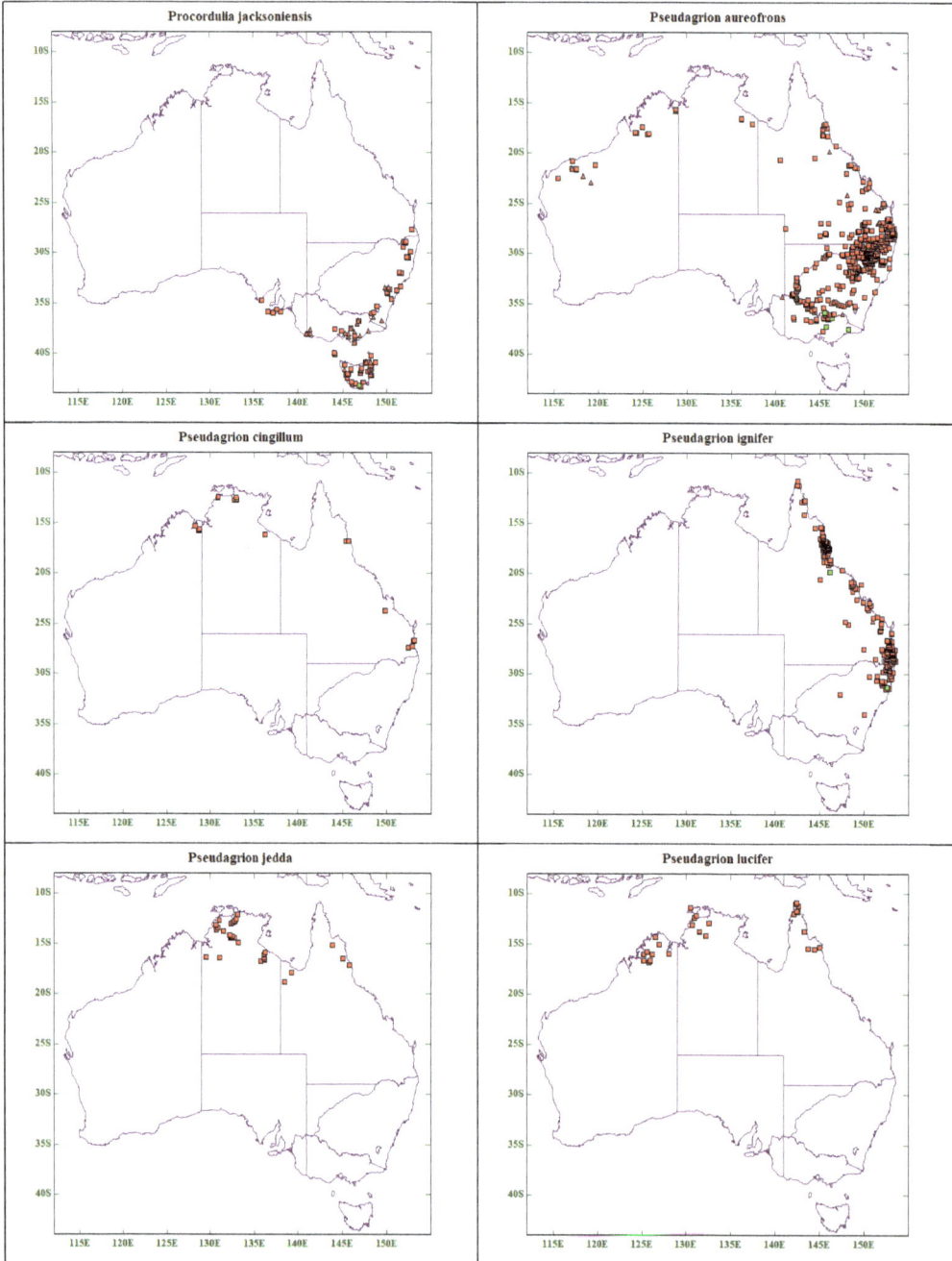

Procordulia jacksoniensis

Pseudagrion aureofrons

Pseudagrion cingillum

Pseudagrion ignifer

Pseudagrion jedda

Pseudagrion lucifer

Pseudagrion microcephalum

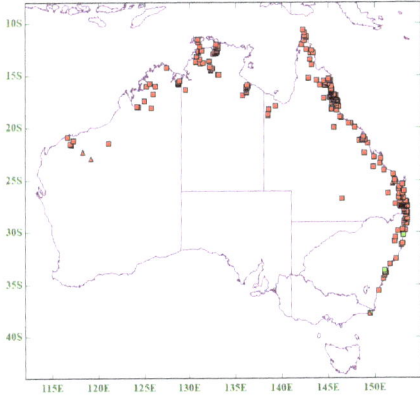

Pseudocordulia circularis

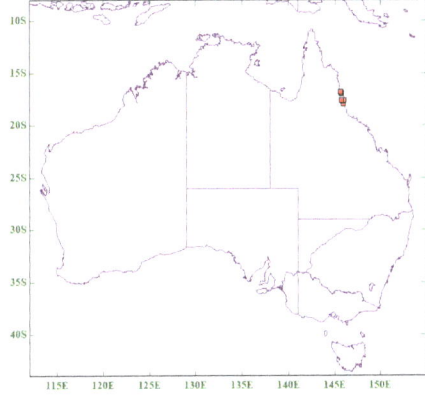

Pseudocordulia elliptica

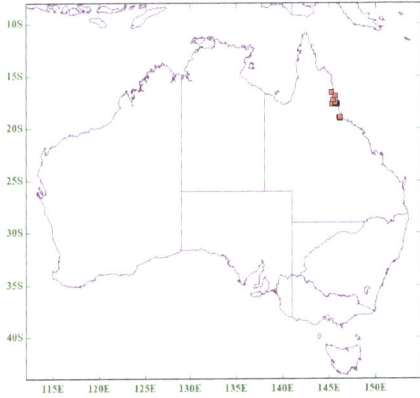

Raphismia bispina

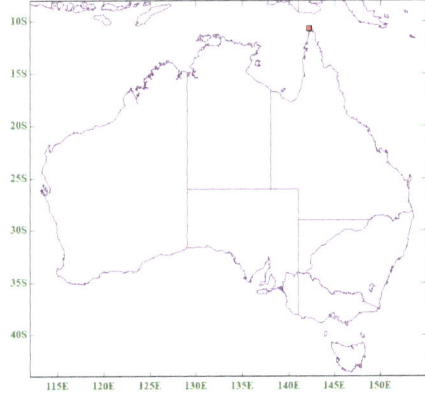

Rhadinosticta banksi

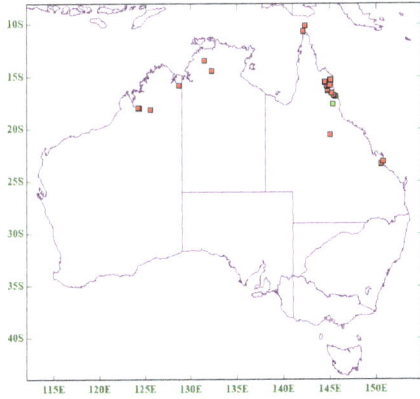

Rhadinosticta simplex

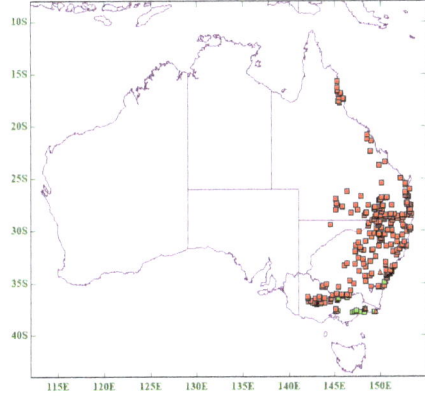

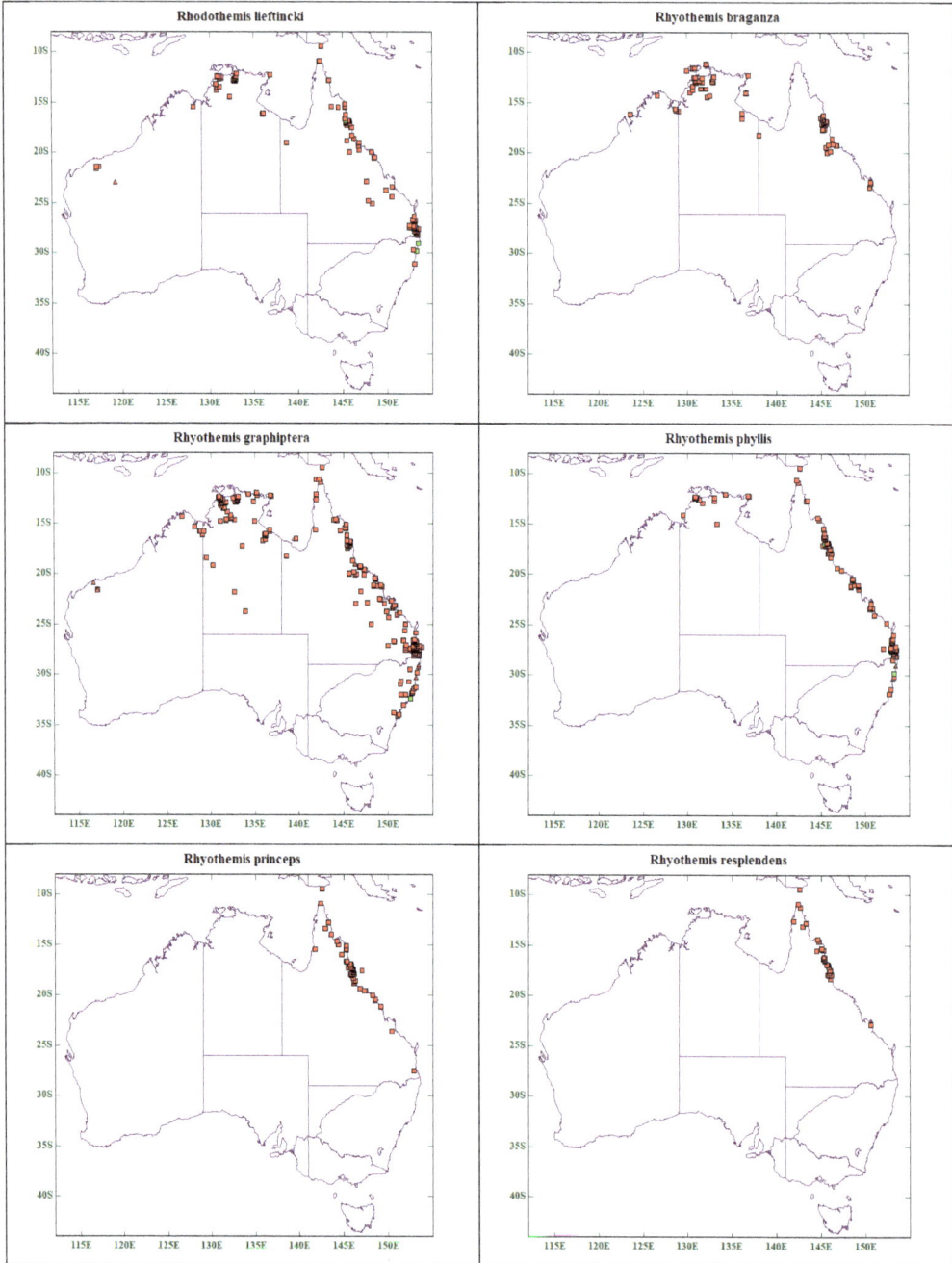

Rhodothemis lieftincki

Rhyothemis braganza

Rhyothemis graphiptera

Rhyothemis phyllis

Rhyothemis princeps

Rhyothemis resplendens

Spinaeschna tripunctata

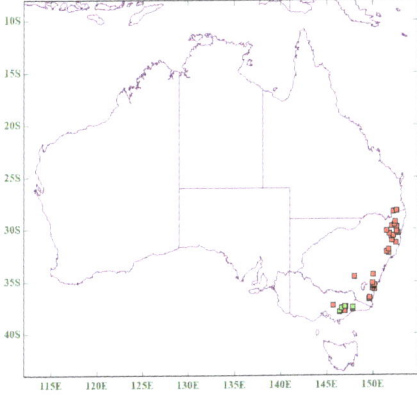

Spinaeschna watsoni

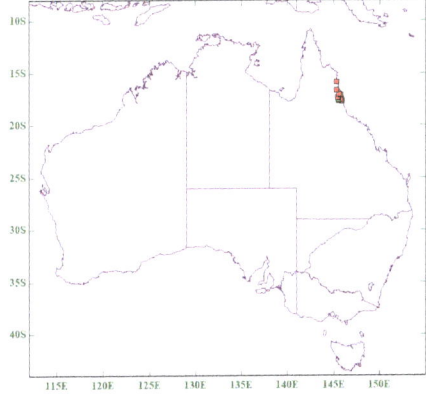

Synlestes selysi

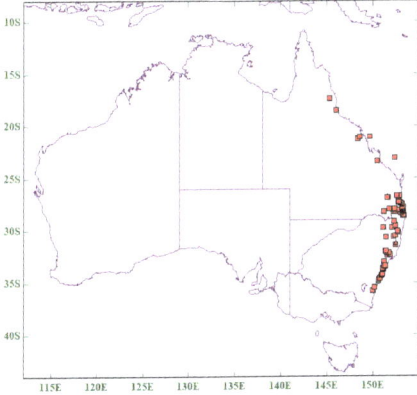

Synlestes tropicus

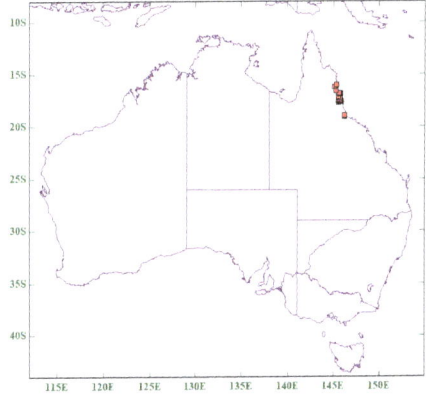

Synlestes weyersii

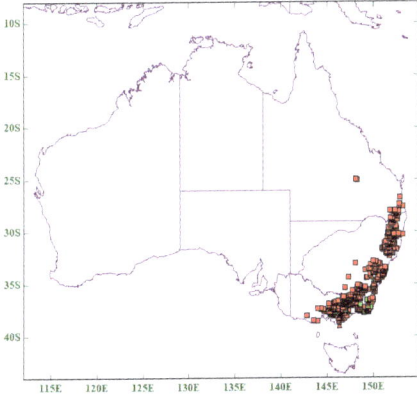

Synthemiopsis gomphomacromioides

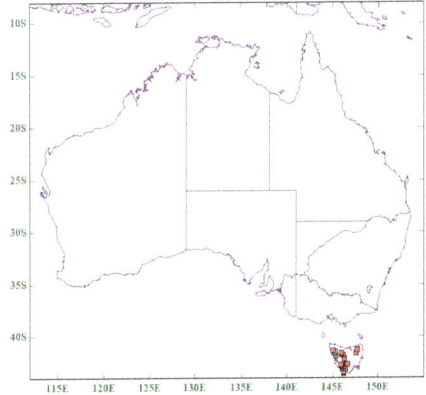

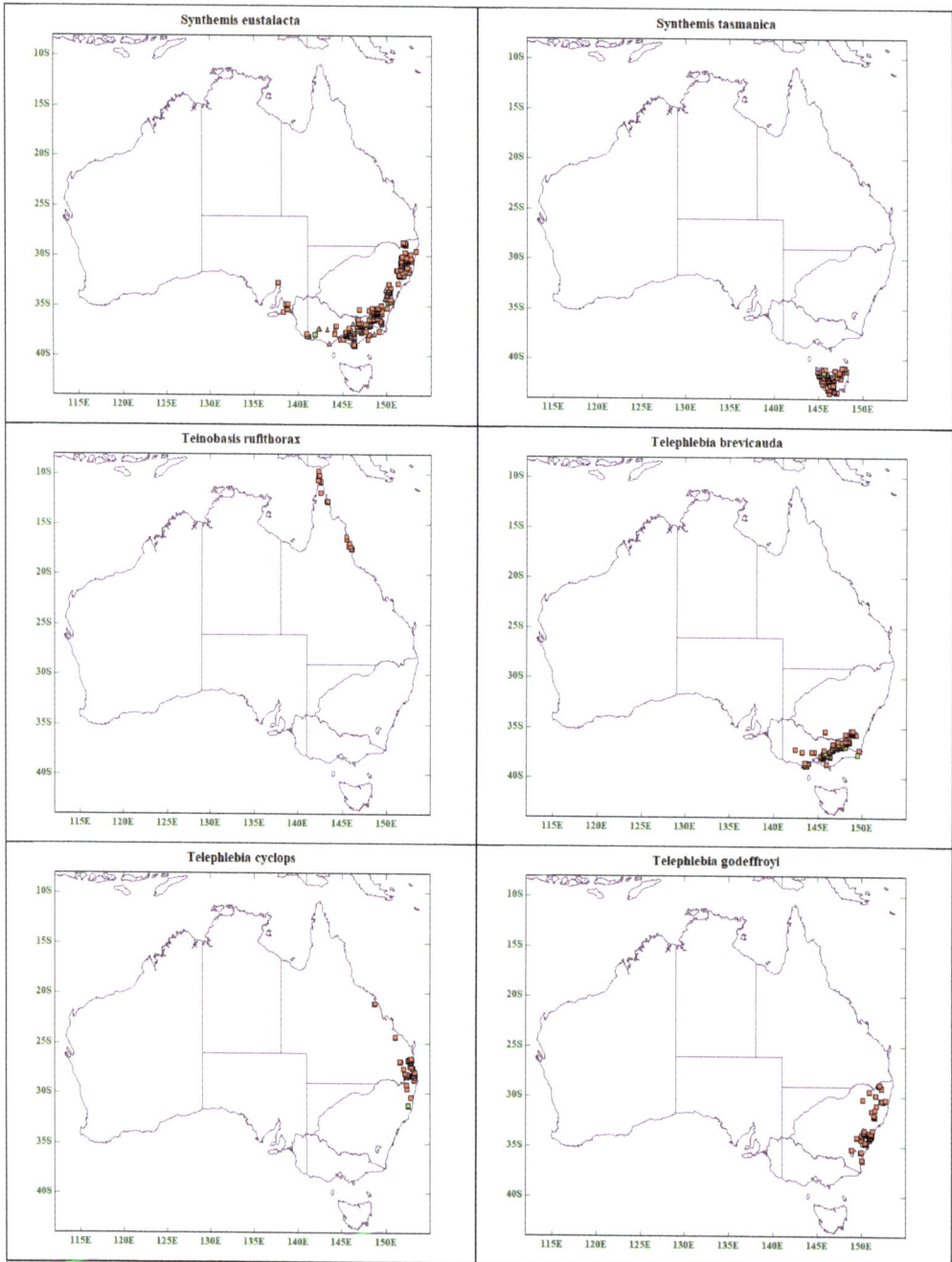

Synthemis eustalacta

Synthemis tasmanica

Teinobasis rufithorax

Telephlebia brevicauda

Telephlebia cyclops

Telephlebia godeffroyi

Telephlebia tillyardi

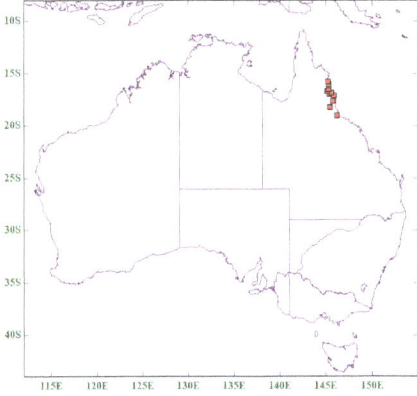

Telephlebia tryoni

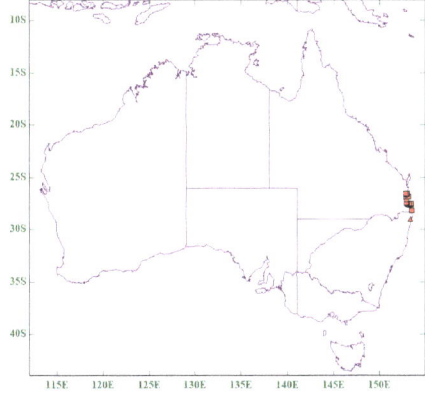

Telephlebia undia

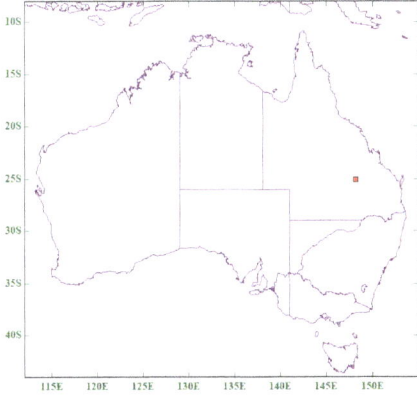

Tetrathemis irregularis cladophylla

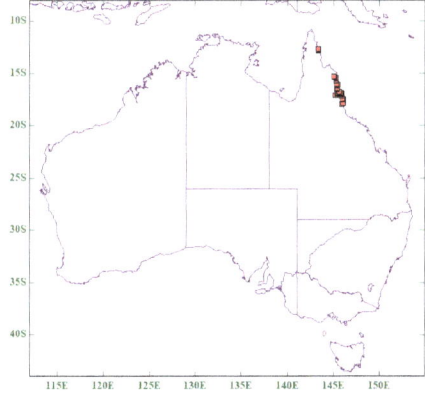

Tholymis tillarga

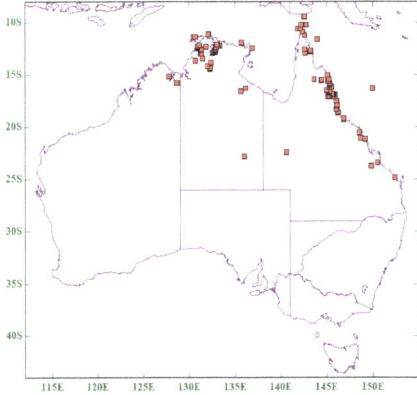

Tonyosynthemis claviculata

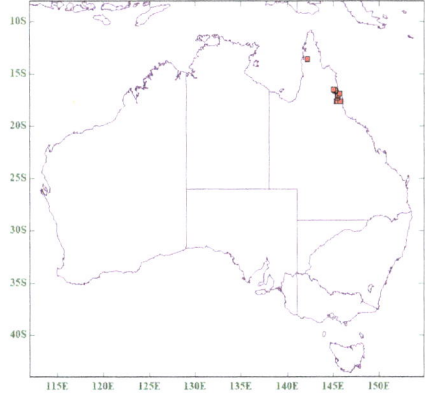

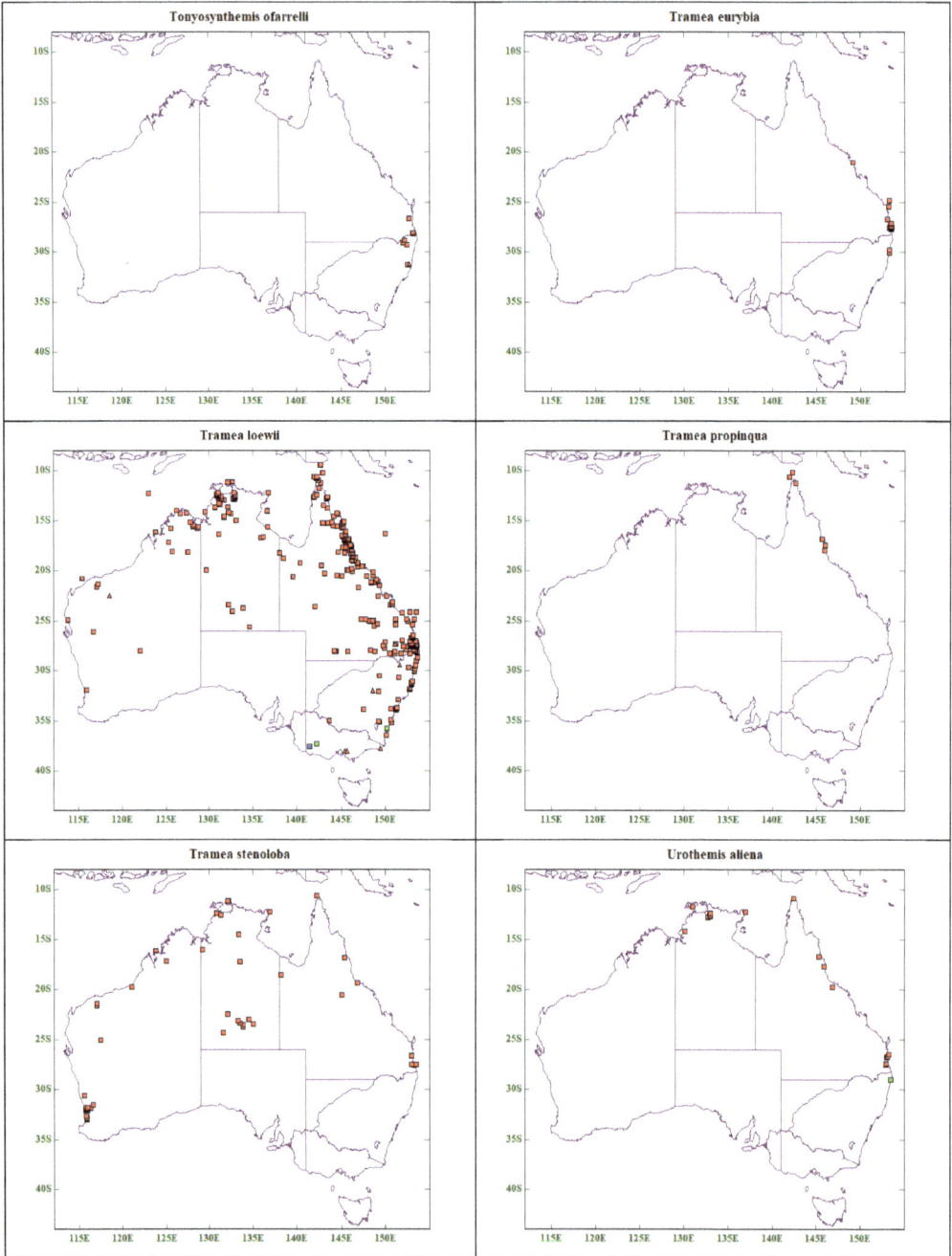

Tonyosynthemis ofarrelli

Tramea eurybia

Tramea loewii

Tramea propinqua

Tramea stenoloba

Urothemis aliena

Xanthagrion erythroneurum

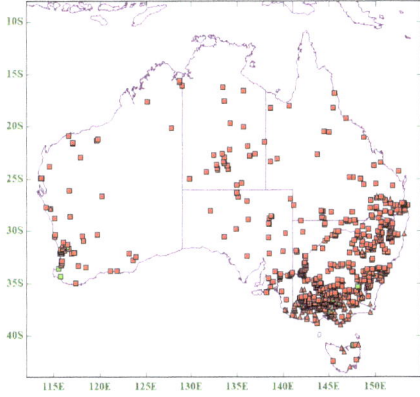

Zephyrogomphus lateralis

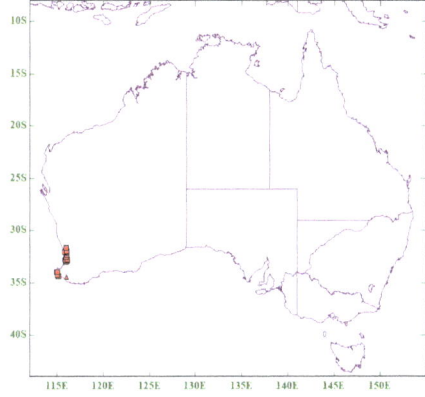

Zephyrogomphus longipositor

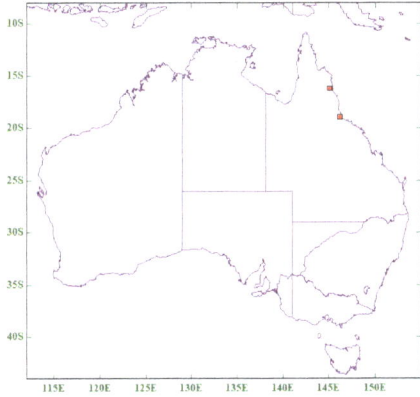

Zyxomma elgneri

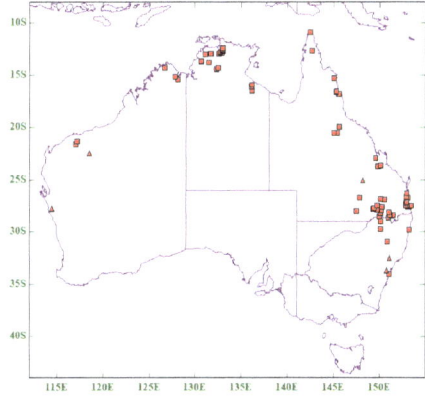

Zyxomma multinervorum

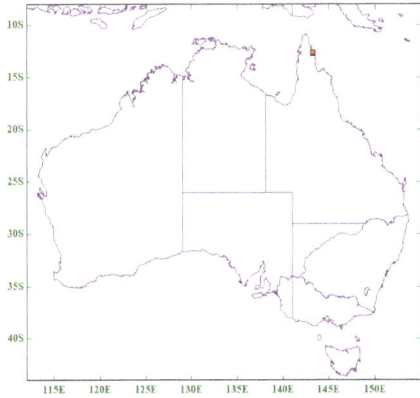

Zyxomma petiolatum

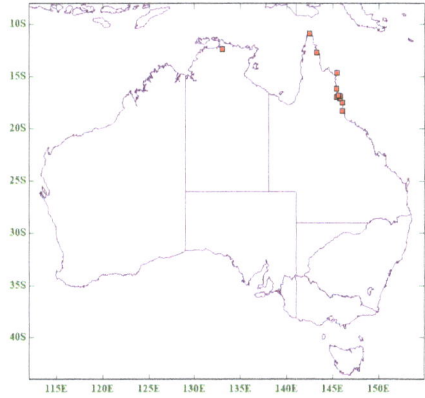

Interim Biogeographical Regions of Australia [IBRA7]

ARC	Arnhem Coast	213	KIN	King	226
ARP	Arnhem Plateau	210	LSD	Little Sandy Desert	71
AUA	Australian Alps	146	MAC	MacDonnell Ranges	223
AVW	Avon Wheatbelt	169	MAL	Mallee	72
BBN	Brigalow Belt North	96	MDD	Murray Darling Depression	128
BBS	Brigalow Belt South	107	MGD	Mitchell Grass Downs	93
BEL	Ben Lomond	232	MII	Mount Isa Inlier	91
BHC	Broken Hill Complex	71	MUL	Mulga Lands	105
BRT	Burt Plain	222	MUR	Murchison	181
CAR	Carnarvon	183	NAN	Nandewar	118
CEA	Central Arnhem	71	NCP	Naracoorte Coastal Plain	156
CEK	Central Kimberley	189	NET	New England Tablelands	121
CER	Central Ranges	72	NNC	NSW North Coast	124
CHC	Channel Country	103	NOK	Northern Kimberley	191
CMC	Central Mackay Coast	100	NSS	NSW South Western Slopes	137
COO	Coolgardie	168	NUL	Nullarbor	71
COP	Cobar Peneplain	130	OVP	Ord Victoria Plain	196
COS	Coral Sea	72	PCK	Pine Creek	207
CYP	Cape York Peninsula	74	PIL	Pilbara	185
DAB	Daly Basin	198	PSI	Pacific Subtropical Islands	71
DAC	Darwin Coastal	201	RIV	Riverina	135

DAL	Dampierland	187		SCP	South East Coastal Plain	151
DEU	Desert Uplands	94		SEC	South East Corner	143
DMR	Davenport Murchison Ranges	72		SEH	South Eastern Highlands	140
DRP	Darling Riverine Plains	116		SEQ	South Eastern Queensland	111
EIU	Einasleigh Uplands	81		SSD	Simpson Strzelecki Dunefields	164
ESP	Esperance Plains	166		STP	Stony Plains	72
EYB	Eyre Yorke Block	165		STU	Sturt Plateau	220
FIN	Finke	225		SVP	Southern Volcanic Plain	154
FLB	Flinders Lofty Block	160		SWA	Swan Coastal Plain	175
FUR	Furneaux	228		SYB	Sydney Basin	131
GAS	Gascoyne	71		TAN	Tanami	72
GAW	Gawler	162		TCH	Tasmanian Central Highlands	230
GES	Geraldton Sandplains	177		TIW	Tiwi Cobourg	204
GFU	Gulf Fall and Uplands	218		TNM	Tasmanian Northern Midlands	234
GID	Gibson Desert	71		TNS	Tasmanian Northern Slopes	72
GSD	Great Sandy Desert	72		TSE	Tasmanian South East	236
GUC	Gulf Coastal	216		TSR	Tasmanian Southern Ranges	238
GUP	Gulf Plains	79		TWE	Tasmanian West	240
GVD	Great Victoria Desert	72		VIB	Victoria Bonaparte	193
HAM	Hampton	71		VIM	Victorian Midlands	148
ITI	Indian Tropical Islands	72		WAR	Warren	173
JAF	Jarrah Forest	171		WET	Wet Tropics	86
KAN	Kanmantoo	158		YAL	Yalgoo	179

Five or less species were recorded for these eight bioregions (and only 47 specimens or sightings) so no chart of "flight" times is given.

	CEA	HAM	GID	LSD	PSI	NUL	BHC	GAS
Lestidae								
Austrolestes annulosus						1		
Coenagrionidae								
Agriocnemis exsudans					4			
Ischnura aurora					7			
Ischnura heterosticta							1	
Xanthagrion microcephalum						1	3	
Aeshnidae								
Hemianax papuensis		2				1	1	
Corduliidae								
Hemicordulia australiae					7			
Hemicordulia tau			1			1	4	
Libellulidae								
Diplacodes bipunctata			2	1				2
Diplacodes haematodes							2	1
Orthetrum caledonicum				1				1
Pantala flavescens								1
Rhyothemis graphiptera	1							
Tramea stenoloba								1
Number of Species	1	1	2	2	3	4	5	5
Number of Specimens	1	2	3	2	18	4	11	6

BHC   Broken Hill Complex
CEA   Central Arnhem
GAS   Gascoyne
GID   Gibson Desert
HAM   Hampton
LSD   Little Sandy Desert
NUL   Nullarbor
PSI   Pacific Subtropical Islands

Between six and ten species were recorded for these ten bioregions so no chart of "flight" times is given.

	COS	ITI	DMR	CER	GSD	GVD	MAL	STP	TAN	TNS
Lestidae										
Austrolestes analis							2			4
Austrolestes annulosus							7	1		2
Austrolestes aridus			7	4	6					
Isostictidae										
Austrosticta fieldi			1							
Coenagrionidae										
Ischnura aurora		1	3	1				18		1
Ischnura heterosticta						1		9		
Xanthagrion microcephalum			2	1	2	3	1	8	2	1
Aeshnidae										
Adversaeschna brevistyla						1	1			
Anax guttatus		1								
Austroaeschna hardyi										16
Austroaeschna inermis										
Austroaeschna ingrid										
Austroaeschna muelleri										
Austroaeschna multipunctata										
Austroaeschna obscura										
Austroaeschna parvistigma										12
Austroaeschna tasmanica										13
Austroaeschna unicornis										5
Hemianax papuensis	4	2		5	2	3	2	5	3	
Synthemistidae										
Synthemis tasmanica										5
Corduliidae										
Hemicordulia tau				2	2	2	7	5	3	4
Libellulidae										
Diplacodes bipunctata		1	3		2	1	1	12	8	
Diplacodes haematodes			6	1	4	2		6	5	
Lathrecista asiatica festa	1									
Macrodiplax cora									1	
Nannophya occidentalis							2			

	1	2	3	4	5	6	7	8	9	10
Neurothemis stigmatizans	1									
Orthetrum caledonicum			4	3	4	3	2	16	12	
Pantala flavescens	1	1		1					2	
Rhyothemis graphiptera									1	
Tholymis tillarga	2									
Tramea loewii	1	2							1	
Tramea stenoloba				2						
Number of Species	6	7	7	8	8	8	9	9	10	10
Number of Specimens	10	9	26	18	24	16	25	80	38	63

CER   Central Ranges

COS   Coral Sea

DMR   Davenport Murchison Ranges

GSD   Great Sandy Desert

GVD   Great Victoria Desert

ITI   Indian Tropical Islands

MAL   Mallee

STP   Stony Plains

TAN   Tanami

TNS   Tasmanian Northern Slopes

Collection dates of specimens of larvae cannot contribute to flight calendars for adults so they have been ignored. However, on the rare occasion where a larva species has been sampled without a comparable adult being recorded, that larva is included in the total number of species.

Not all museum labels record the date of capture. The number of specimens without a date has been included in the final column headed "n.d.". This allows the full number of records on file to be seen so as to give some idea of which are the predominant species in the zone.

The order of the IBRA7 zones is not alphabetical but approximately contiguous starting at Cape York Peninsula. An index is given on page 69.

CYP   Cape York Peninsula   12,256,457 ha

Complex geology dominated by the Torres Strait Volcanics in the north, the metamorphic rocks and acid intrusive rocks of various ages of the Coen-Yambo Inlier which runs north- south along the eastern margin of the region and encompasses the high-altitude/high-rainfall areas of Iron Range and McIlwraith Range. The deeply dissected sandstone plateaus and ranges of the Battle Camp Sandstones lie in the south of the region adjacent to the undulating Laura Lowlands composed of residual weathered sands and flat plains of colluvial and alluvial clays, silts and sands. The west of the region is dominated in the south by the extensive Tertiary sand sheet dissected by intricate drainage systems of the Holroyd Plain, the Tertiary laterite of the undulating Weipa Plateau, the low rises of Mesozoic sandstones, with the northern extension of the Weipa Plateau and extensive coastal plains adjoining the Gulf of Carpentaria. Extensive aeolian dune fields lie in the east associated with Cape Bedford/Cape Flattery in the south and the Olive and Jardine Rivers.

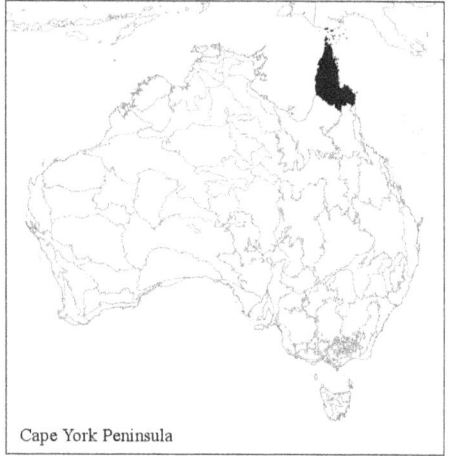

Cape York Peninsula

The vegetation is predominantly *Eucalyptus tetrodonta* and *Corymbia tessellaris/C. clarksoniana* woodlands, *Melaleuca viridiflora* woodlands, heathlands and sedgelands, notophyll vine forests, with semi-deciduous mesophyll vine forests on the eastern ranges and deciduous vine thickets on drier western slopes. Extensive mangrove forests are found in Kennedy Inlet in the north east of the region and estuaries on both the west and east coasts. Tropical humid/maritime climate, with rainfall varying from 1000 mm to 1600 mm .

Species   117   Specimens   2631   Adults   2631   Larvae   0

	J	F	M	A	M	J	J	A	S	O	N	D	n.d.
Family Lestidae													
Austrolestes insularis	2	4	28		1	1	10	3		3			
Indolestes alleni	2				1		2	1		1	1	2	
Indolestes tenuissimus	4	2					1			2		12	1
Lestes concinnus	3	1	9		3		3			2	2	9	
Family Argiolestidae													
Austroargiolestes icteromelas										1			
Family Lestoideidae													
Diphlebia euphoeoides	1	1					1	1		2			1
Family Isostictidae													
Eurysticta reevesi			4										
Labidiosticta vallisi										1			
Oristicta filicicola	1	1	8	15	1	2				4	11	2	
Rhadinosticta banksi		2	10										
Rhadinosticta simplex	1												
Family Platycnemidae													
Nososticta coelestina	4		41			10	1	3	5		6	7	
Nososticta solida	1	1	2										
Nososticta solitaria	3	3	12			1	6		4	3	28	5	
Family Coenagrionidae													
Aciagrion fragile	1	1	4	4		1	1	1	4	5	19	2	3
Agriocnemis femina				2									
Agriocnemis pygmaea	1		2				5		1	1			
Agriocnemis rubricauda	1						7		35	22	1		
Archibasis mimetes	4					1	1						
Argiocnemis rubescens			1	1		2	2		21	20			
Austroagrion exclamationis			11	2	4		11	6	2	21	23	7	
Austrocnemis maccullochi										31			
Ceriagrion aeruginosum	3						3		4	6	4	4	
Ischnura aurora			4		6	2	13		2	16	23	5	1
Ischnura heterosticta	4	5	5	12	2	1	9		1	14	38	9	1
Ischnura pruinescens			2				5			8	4	2	5
Pseudagrion ignifer			1	4		4			5	9		3	
Pseudagrion jedda										1			
Pseudagrion lucifer		1	19			7	1	1			3		
Pseudagrion microcephalum	3	1	9		14		3	1	3	11			
Teinobasis rufithorax	11	4	8	22		4	4				1	4	

	J	F	M	A	M	J	J	A	S	O	N	D	n.d.
Family Aeshnidae													
Agyrtacantha dirupta	1					3	3				1		
Anaciaeschna jaspidea		2	1		2								1
Anax gibbosulus	3	1	1				1				1		1
Anax guttatus	10	2									2	15	
Austrogynacantha heterogena	3	1	1		2	1							
Austrophlebia costalis													
Austrophlebia subcostalis													
Dendroaeschna conspersa													
Dromaeschna forcipata	1		1								2		
Dromaeschna weiskei													
Gynacantha dobsoni			1		1	21	10	1	1	3	2		
Gynacantha kirbyi			1			3		1		2	1		
Gynacantha mocsaryi	4		3	2		5				1	1	3	1
Gynacantha nourlangie					3	2	4				1		
Gynacantha rosenbergi	3	3				1					8		
Hemianax papuensis	1		9	1	1	1	3	1		1	3		
Family Petaluridae													
Petalura pulcherrima	1	4										2	1
Family Gomphidae													
Antipodogomphus edentulus	1										3		
Antipodogomphus proselythus	4	2									6		2
Austroepigomphus turneri	1									2	2		
Austrogomphus arbustorum										1	2		2
Austrogomphus bifurcatus	1												
Austrogomphus divaricatus		1								1			
Austrogomphus mjobergi	3		2	1									
Austrogomphus prasinus	3	1	2	6	1						2	5	7
Hemigomphus comitatus		2									2	2	
Ictinogomphus australis	2	1									3	3	3
Ictinogomphus paulini	2									2	1		
Family Synthemistidae													
Choristhemis flavoterminata		1	1										
Eusynthemis nigra											3		
Tonyosynthemis claviculata											1		
Family Macromiidae													
Macromia tillyardi			4							1			1
Macromia viridescens	1			1							1		1

	J	F	M	A	M	J	J	A	S	O	N	D	n.d.
Family Corduliidae													
Hemicordulia continentalis	1	2	1										
Hemicordulia intermedia	1	1	4		1				1				
Hemicordulia kalliste	1												
Pentathemis membranulata	3											1	
Family Libellulidae													
Aethriamanta nymphaea	1												
Agrionoptera insignis	2	3	6	3		2	1	1			1	2	
Agrionoptera longitudinalis	4		2	1					2			2	
Brachydiplax denticauda	6	1				1		1	3	2	8	1	1
Brachydiplax duivenbodei	2											1	
Camacinia othello	3	3			1			1	1		1		2
Crocothemis nigrifrons									2	1	1	1	
Diplacodes bipunctata	1	1	20	1	1	2	2						
Diplacodes haematodes	7	1	16	2	13	9	27	4	17	7	8	4	
Diplacodes nebulosa	4		14	1		1	4	2		1	1	4	
Diplacodes trivialis	6	2	4	2	2	5	14			8	2		
Hydrobasileus brevistylus											1	2	
Lathrecista asiatica festa	17	6	8	2	5	11	4	2	14	4	7	9	1
Macrodiplax cora	2		2		2		2				2	3	
Nannodiplax rubra	8		22	2	3	8	32	25	6	7	8	4	1
Nannophlebia eludens	2		10	4		4	3			1	4		
Nannophlebia risi	1		2					1	2		1		
Nannophya australis	1		7		1				1				
Nannophya paulson	2												
Neurothemis oligoneura	3	9		1	1	12	4	9	14	3	1	3	5
Neurothemis stigmatizans	15	7	17	7	12	33	26	3	20	8	11	19	7
Notolibellula bicolor		3											1
Orthetrum balteatum	2					1		1					
Orthetrum caledonicum	2		5	1	1		7	2		5	2	5	1
Orthetrum migratum	5	3	17										
Orthetrum sabina	6	1	3	1	1		2	2	4	1	2	4	1
Orthetrum serapia	2			3	6	4	6	3	2	3	6		3
Orthetrum villosovittatum	6	4	3	2		3		3	6	5	4	4	2
Pantala flavescens	8	8	14		3	1	5					1	1
Potamarcha congener	1		2	1			6		2		1		
Raphismia bispina					4								
Rhodothemis lieftincki	5	1	2				1	1				6	

	J	F	M	A	M	J	J	A	S	O	N	D	n.d.
Rhyothemis graphiptera	11					2	1		1	5		4	
Rhyothemis phyllis	8		1							3	3	7	1
Rhyothemis princeps	4	1		3	1						2	2	
Rhyothemis resplendens	10	2	2								3	10	
Tetrathemis irregularis	2	2	4			1	1			1	1	3	
Tholymis tillarga	4	2	9	1		11	4		1		4	8	
Tramea loewii	5	4	11	2	3	3	9	1		3	3	3	2
Tramea propinqua			2				1						
Tramea stenoloba	1											2	
Urothemis aliena	5												
Zyxomma elgneri			1				1					4	
Zyxomma multinervorum	2												
Zyxomma petiolatum						1						2	
Genera incertae sedis													
Austrocordulia refracta		1											
Micromidia·atrifrons	1					2							
Micromidia rodericki	1												

GUP   Gulf Plains   22,041,825 ha

Marine and terrestrial
deposits of the Carpentaria
and Karumba basins; plains,
plateaus and outwash plains;
woodlands and grasslands.

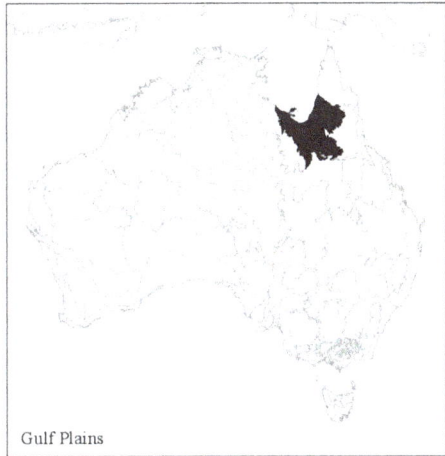

Gulf Plains

Species   36   Specimens   157   Adults   157   Larvae   0

	J	F	M	A	M	J	J	A	S	O	N	D	n.d
Family Lestidae													
Lestes concinnus				3	3					1			
Family Isostictidae													
Austrosticta fieldi				4									
Austrosticta frater				11									
Family Platycnemidae													
Nososticta fraterna				2									
Family Coenagrionidae													
Agriocnemis argentea							1						
Argiocnemis rubescens				5									
Austroagrion exclamationis												1	
Austroagrion watsoni				4									
Austrocnemis maccullochi												1	
Ischnura aurora				8	1					1		1	
Ischnura heterosticta				1	1				1				
Ischnura pruinescens				1									
Pseudagrion jedda							2						
Pseudagrion microcephalum				3			1						
Xanthagrion erythroneurum				1								1	

	J	F	M	A	M	J	J	A	S	O	N	D	n.d
Family Aeshnidae													
Adversaeschna brevistyla									1	1			
Anax guttatus	3												
Austrogynacantha heterogena				1									
Gynacantha nourlangie				4									
Hemianax papuensis	1		1	2	2								1
Family Corduliidae													
Hemicordulia australiae									1	2			
Hemicordulia tau				1									
Family Libellulidae													
Diplacodes bipunctata			3	6	19	2	1				1		
Diplacodes haematodes			7				1						
Nannodiplax rubra			8										
Nannophya australis		1											
Neurothemis stigmatizans				1	5	2							
Notolibellula bicolor				2									
Orthetrum caledonicum				1	2								
Orthetrum migratum													1
Pantala flavescens				1	2	1			1				
Potamarcha congener			1										
Rhyothemis graphiptera					2	1							
Rhyothemis princeps			1										
Rhyothemis resplendens													
Tramea loewii				1					2				1

EIU   Einasleigh Uplands   11,625,726 ha

High plateau of Palaeozoic sediments, granites, and basalts; dominated by ironbark (*Eucalyptus* spp.) woodlands.

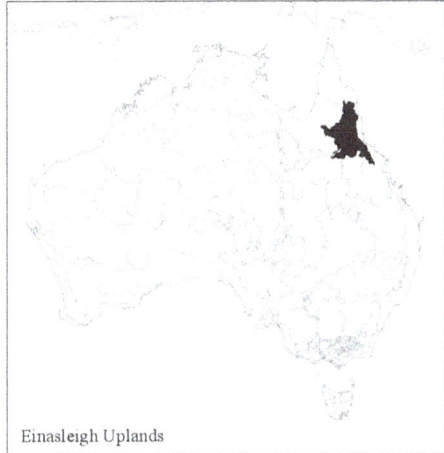

Einasleigh Uplands

Species   120   Specimens   1491   Adults 1491   Larvae   0

	J	F	M	A	M	J	J	A	S	O	N	D	n.d.
Family Synlestidae													
Synlestes selysi												1	
Family Lestidae													
Austrolestes insularis	6		3	4	3				1			2	
Austrolestes leda	5								1		1		
Indolestes alleni	1												
Indolestes tenuissimus	2										1		
Lestes concinnus			1	6	3	1	2			1		1	
Family Argiolestidae													
Austroargiolestes aureus												1	
Austroargiolestes icteromelas	3			8				1	2		1		
Family Lestoideidae													
Diphlebia euphoeoides	1	1	7	4	6						6	15	
Lestoidea lewisiana												1	
Family Isostictidae													
Austrosticta frater		4	14	1									
Eurysticta reevesi				1									
Neosticta fraseri		1										1	
Oristicta filicicola					1					1		1	

	J	F	M	A	M	J	J	A	S	O	N	D	n.d.
Oristicta rosendaleorum			6										
Rhadinosticta banksi		11	16	1	1								
Rhadinosticta simplex			1		1						1		
Family Platycnemidae													
Nososticta coelestina	1		2	7	3						2	8	
Nososticta solida	2	1	11	2					5	7	3	4	
Nososticta solitaria		5	19	9	3					7	8	22	
Family Coenagrionidae													
Aciagrion fragile	8			8							1		
Agriocnemis argentea		2	1	5	2					1	1	5	
Agriocnemis pygmaea	1	1		6	1				1	1	3	2	
Agriocnemis rubricauda			2		1						1	1	
Argiocnemis rubescens			2	1	3	1			6		1	3	
Austroagrion exclamationis		2		1	1					1	2	4	
Austroagrion watsoni			1	1					3	2	1	8	
Austrocnemis maccullochi												1	
Austrocnemis splendida				1							2	3	
Ceriagrion aeruginosum	4	4		1	1						1	2	
Ischnura aurora	1		6		3		2		4	5	6	6	
Ischnura heterosticta	2	5	2	3	1				3	2	3	15	
Ischnura pruinescens		2	1	3					3		5	2	
Pseudagrion aureofrons			2						1	1		1	
Pseudagrion cingillum												2	
Pseudagrion ignifer	9		3	3	6			1	1	2	10	18	
Pseudagrion jedda												2	
Pseudagrion microcephalum			7	3	1				4	2	13	12	1
Xanthagrion erythroneurum									1	1			
Family Aeshnidae													
Adversaeschna brevistyla											1		
Anax gibbosulus		2	1									2	
Anax guttatus		1			1							1	
Austroaeschna speciosa		1									1	3	
Austrogynacantha heterogena		2	2	1					1		1		
Dromaeschna forcipata	3	2									8	4	
Dromaeschna weiskei	1										1		
Gynacantha dobsoni										1	2		
Gynacantha mocsaryi	1												

82

	J	F	M	A	M	J	J	A	S	O	N	D	n.d.
Gynacantha nourlangie								1		1	1		
Gynacantha rosenbergi										1	3		
Hemianax papuensis		2		1								2	
Spinaeschna watsoni		2										1	
Telephlebia tillyardi	1		1								6	1	
Family Petaluridae													
Petalura ingentissima	2	1											
Family Gomphidae													
Antipodogomphus acolythus				1									
Antipodogomphus neophytus				1									
Antipodogomphus proselythus												2	
Austroepigomphus turneri		1									3	9	
Austrogomphus amphiclitus	1		3	1						2		5	1
Austrogomphus arbustorum			1								1	3	
Austrogomphus bifurcatus			2									2	
Austrogomphus cornutus										1			
Austrogomphus divaricatus		1	2								5	5	
Austrogomphus doddi	2		1	1									
Austrogomphus mjobergi	1	1									1		
Austrogomphus prasinus	3	1	1	8							5	6	
Hemigomphus comitatus	5			1						1	6	13	
Hemigomphus heteroclytus												1	
Hemigomphus magela													
Hemigomphus theischingeri			1										
Ictinogomphus australis		3		1						1	4	15	
Family Synthemistidae													
Choristhemis flavoterminata	4			5							10	13	
Eusynthemis nigra	2	1								2	2	1	
Tonyosynthemis claviculata	1	1	2	1									
Family Macromiidae													
Macromia tillyardi	1										1	1	
Family Corduliidae													
Hemicordulia australiae		1									1	1	
Hemicordulia continentalis		5										1	
Hemicordulia intermedia	3	1	2	2						1	5	9	
Hemicordulia tau			1									1	

	J	F	M	A	M	J	J	A	S	O	N	D	n.d.
Pentathemis membranulata											4	4	
Family Libellulidae													
Aethriamanta circumsignata												7	
Agrionoptera longitudinalis											1		
Brachydiplax denticauda			2							3	2	1	
Crocothemis nigrifrons		4	2	2	3					1	1	8	
Diplacodes bipunctata	2	3	7	4	6		2		10	8	3	8	
Diplacodes haematodes	4	3	19	6	18	1	1		1	11	3	31	
Diplacodes nebulosa		1		3			3	1			1	2	
Diplacodes trivialis	2	9	2	2	5				6		4	16	
Hydrobasileus brevistylus	2	5	3	1								7	
Lathrecista asiatica festa		1	1	2	1						2	3	
Macrodiplax cora		1			1							3	
Nannodiplax rubra	1				5					7	7		
Nannophlebia eludens	1	1	1	10	2						1		
Nannophlebia risi	1	1	5	1	1					2	2	1	
Nannophya australis											1		
Neurothemis stigmatizans	1	1	5	3	8						4	4	
Orthetrum caledonicum	3	3	3	3	4					6		11	
Orthetrum migratum		1	4	1							4	5	
Orthetrum sabina		3		1	2		1			2		4	
Orthetrum serapia		1											
Orthetrum villosovittatum	2	6	2	3	5					1	2	7	
Pantala flavescens		2	4	1						4	1	2	
Potamarcha congener												1	
Rhodothemis lieftincki				1	1						2	1	
Rhyothemis braganza	3	1		1	1					1	6	8	
Rhyothemis graphiptera	1	3	1	2					1	1	1	9	
Rhyothemis phyllis				2							1		
Rhyothemis princeps	2		1										
Tetrathemis i.cladophila										3			
Tholymis tillarga		1									1	4	
Tramea loewii	2	2	3		2		2			2	1	7	
Tramea stenoloba												1	
Zyxomma elgneri				1								1	
Zyxomma petiolatum											1		

	J	F	M	A	M	J	J	A	S	O	N	D	n.d.
Genera incertae sedis													
Archaeophya magnifica		2											
Austrocordulia refracta											1	2	
Austrophya mystica		1									2	2	
Micromidia·atrifrons		9											
Pseudocordulia elliptica											1		

WET   Wet Tropics   1,989,107 ha

The bioregion is dominated by rugged rainforested mountains, including the highest in Queensland Mt Bartle Frere (1622m). It also includes extensive plateau areas along its western margin, as well as low lying coastal plains. The most extensive lowlands are in the south, associated with the floodplains of the Tully and Herbert Rivers. Most of the bioregion drains to the Coral Sea from small coastal catchments, but higher western areas drain in the south into the Burdekin River, and in the north into tributaries of the Mitchell River. The region contains extensive areas of tropical rainforest, plus beach scrub, tall open forest, open forest, mangrove and Melaleuca woodland communities.

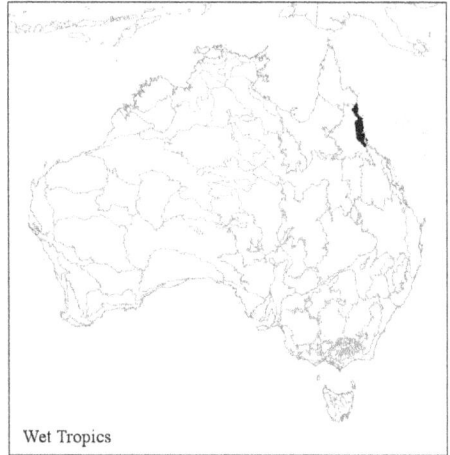

Wet Tropics

Species   145   Specimens   5098   Adults   5098   Larvae   0

	J	F	M	A	M	J	J	A	S	O	N	D	n.d.
Family Synlestidae													
Chorismagrion risi	3	2	7	9	4		2		1	7	12	1	3
Episynlestes cristatus	10	3	5	9							4		1
Synlestes selysi			8	1									
Synlestes tropicus	4	5	9	6	10		1			6	13	11	3
Family Lestidae													
Austrolestes insularis	1												
Austrolestes leda											1		
Indolestes alleni	5			2			1	1		3		1	
Indolestes tenuissimus	5		2	2					5	2	4	6	1
Lestes concinnus	1		2	3	1	1		3			1		
Family Argiolestidae													
Austroargiolestes aureus	10	10	13	1				1	5	21	27	33	4
Austroargiolestes icteromelas	5	1	12	9				1	2	3	2	8	1
Griseargiolestes metallicus	1		1						5	7	3	4	

	J	F	M	A	M	J	J	A	S	O	N	D	n.d.
Podopteryx selysi	6	1	4	5					3	2	3		
Family Lestoideidae													
Diphlebia euphoeoides	39	12	71	51	4	1		11	50	24	37	40	1
Diphlebia hybridoides	5	2							3	17	29		
Lestoidea barbarae											10	4	
Lestoidea brevicauda	1		2							4	5		
Lestoidea conjuncta	23	5	11	9	1	2			4	13	24	11	2
Lestoidea lewisiana											2	11	
Family Isostictidae													
Labidiosticta vallisi			1										
Neosticta fraseri	2		12	3	1				1	1	8	8	
Oristicta filicicola	8	4	10	11				3	2	6	6	4	
Rhadinosticta banksi				1									
Rhadinosticta simplex	1		2	3						3	1	1	
Family Platycnemidae													
Nososticta coelestina	4	3	12	12	2				2	14	6	15	
Nososticta solida	5		6	6	1				12	6	10	7	
Nososticta solitaria	15	8	41	18	1		1	5	12	25	19	48	2
Family Coenagrionidae													
Aciagrion fragile	2		1	5		1	1	4	4	8	12	6	
Agriocnemis argentea	3		6		2					7	11	12	1
Agriocnemis dobsoni	3								2	3	8		
Agriocnemis pygmaea	1	1	2	4	1				10	3	7	3	
Agriocnemis rubricauda										1	1		
Archibasis mimetes	2			5					7	8	2	1	1
Argiocnemis rubescens	1	1	6	5	5	1	1	4	18	12	5	9	8
Austroagrion exclamationis		1				1					1	3	
Austroagrion watsoni			3	1					25		2	6	
Austrocnemis maccullochi												6	
Austrocnemis splendida	3								1		4	1	
Ceriagrion aeruginosum	4		6	3	3		2		9	14	15	13	4
Ischnura aurora	3		14	1	2			3	25		2	6	
Ischnura heterosticta	3		4	5				1	16	2	2	1	
Ischnura pruinescens	10		2	3	1		1		15	3		1	1
Pseudagrion aureofrons			1						1		1	2	1
Pseudagrion cingillum	1								1		1	1	
Pseudagrion ignifer	12	4	24	17	7		2	9	46	24	24	19	6
Pseudagrion jedda	1												

	J	F	M	A	M	J	J	A	S	O	N	D	n.d.
Pseudagrion microcephalum	4	1	11	13	1		1	3	22	11	19	17	2
Teinobasis rufithorax			4	1					1	8		2	
Xanthagrion erythroneurum			1										
Family Aeshnidae													
Adversaeschna brevistyla	1							1			1		
Anax gibbosulus	1	3			1			1	2				
Anax guttatus	10	1									1		2
Austroaeschna speciosa	2	1	2										1
Austrogynacantha heterogena	4	5	2	1						1			
Austrophlebia subcostalis	1							1			3	2	
Dromaeschna forcipata	22	8	7	3						5	16	10	3
Dromaeschna weiskei	10		8	11	2					2	8	6	1
Gynacantha dobsoni	6		2		1			9	2	9	4	1	2
Gynacantha kirbyi				1						1			
Gynacantha mocsaryi	6	1	1						3	3	1	2	
Gynacantha rosenbergi	6	2	2		3				1	14	8	8	5
Hemianax papuensis	6	4	1	2				3	1		1	2	1
Spinaeschna watsoni	1		1						3		2		
Telephlebia tillyardi	1	3	2							3	2	2	
Family Petaluridae													
Petalura ingentissima	19	1	1								1	4	2
Family Gomphidae													
Antipodogomphus acolythus	1											1	
Antipodogomphus proselythus	5	1									5	2	
Austroepigomphus praeruptus				1						1			
Austroepigomphus turneri		3									2		
Austrogomphus amphiclitus	2	5	2	11							1	1	1
Austrogomphus arbustorum	1										4		3
Austrogomphus bifurcatus	3	8	5	5					1	1	3	2	
Austrogomphus cornutus	1											3	
Austrogomphus divaricatus	3	5		3						1	7	2	
Austrogomphus doddi		5							1				1
Austrogomphus prasinus	31	5	12	14					1	6	17	11	
Hemigomphus atratus										1			
Hemigomphus comitatus	9	3	1							2	8	8	
Hemigomphus heteroclytus	1		1								1		
Hemigomphus theischingeri	1		3								6	5	
Ictinogomphus australis	5	5									2	6	3

	J	F	M	A	M	J	J	A	S	O	N	D	n.d.
Odontogomphus donnellyi											3		
Zephyrogomphus longipositor												2	
Family Synthemistidae													
Choristhemis flavoterminata	18	5	10				1		3	5	12	31	
Choristhemis olivei	1											1	
Eusynthemis barbarae												5	
Eusynthemis netta													1
Eusynthemis nigra	13	6	5	3						4	24	26	3
Eusynthemis tenera										1		1	
Tonyosynthemis claviculata	1	6	1								1	1	
Family Macromiidae													
Macromia tillyardi	8				1					4			
Family Corduliidae													
Hemicordulia australiae	2	1	2		1			2	2	1	3	3	3
Hemicordulia continentalis	5	1	2						3	4	5	4	
Hemicordulia intermedia		1	4					1	1		7	2	
Hemicordulia tau	1											1	
Family Libellulidae													
Aethriamanta nymphaea			2							1			
Agrionoptera insignis	3	3	11	15	3	2			1	9	2	6	
Agrionoptera longitudinalis	12	2	11	2	2		1	1	7	3	8	3	
Brachydiplax denticauda	8	1	5	4					5	7	4		1
Brachydiplax duivenbodei	10		14	1					1	1		2	1
Camacinia othello	4	1											
Crocothemis nigrifrons	5			1		2			6	6	4	8	1
Diplacodes bipunctata	28		14	11		3	2	1	11	2	11	4	
Diplacodes haematodes	16		25	14	8	4	2	6	33	7	6	7	1
Diplacodes nebulosa	1		1	1		1			1			1	
Diplacodes trivialis	7	1	22	9	6	3	4	3	6	1	6	3	1
Hydrobasileus brevistylus	4	2							1	5	6	4	
Lathrecista asiatica festa	13		10	7	1			4	3	4	2	7	1
Macrodiplax cora	2	2	2						4	1	6	2	
Nannodiplax rubra	14		7	12	4	6	1	4	13	8	13	10	2
Nannophlebia eludens	11	5	25	17	7	2			6	6	16		2
Nannophlebia risi	1		8	4		1			3	1	7	2	
Nannophya australis	1												
Neurothemis stigmatizans	22	5	42	32	20	26	5	32	26	20	13	34	2
Orthetrum caledonicum	3		5	10	4	4		1	15			4	

	J	F	M	A	M	J	J	A	S	O	N	D	n.d.
Orthetrum migratum	3										1		
Orthetrum sabina	4	1	7	9	4	11	2	11	13	7	5	4	
Orthetrum serapia	6		3	3				3		1	1	1	
Orthetrum villosovittatum	23	3	24	7	6	8	3	5	22	29	16	17	1
Pantala flavescens	10	1	11	18	2	5		8	3		1	2	
Potamarcha congener			1										
Rhodothemis lieftincki	1			3	1	1			2	5	2	4	
Rhyothemis braganza	8								1	1	2	14	
Rhyothemis graphiptera	4		6	2		1			2	3	4	5	2
Rhyothemis phyllis	23	1	5	3	2				9	7	9	4	1
Rhyothemis princeps	13	1	17	3				2	3	9	7	16	1
Rhyothemis resplendens	24	1							2		8	9	1
Tetrathemis irregularis	3	3	5	4	4			1	1	5	3	10	
Tholymis tillarga	6	1	10	2	1	3		2	3	2	3	1	2
Tramea loewii	8		4	11	2	2		11	9	1	3	5	1
Tramea propinqua	1										1	1	
Urothemis aliena											1	1	
Zyxomma elgneri			1	1							1		
Zyxomma petiolatum	5		3	1						1		3	
Genera incertae sedis													
Archaeophya magnifica	2									1	4	6	2
Austrocordulia refracta										2	9	5	
Austrophya mystica	2	1	1							2	2	3	
Cordulephya bidens			2	3	4								
Lathrocordulia garrisoni												2	
Micromidia·atrifrons	3		2									2	
Pseudocordulia circularis	1								1	7	2	1	
Pseudocordulia elliptica	6									1	6	3	

MII   Mount Isa Inlier   6,778,263 ha

Rugged hills and outwash, primarily associated with Proterozoic rocks; skeletal soils; low open eucalypt woodlands dominated by *Eucalyptus leucophloia* and *E.pruinosa*, with a *Triodia pungens* understorey. Semi-Arid.

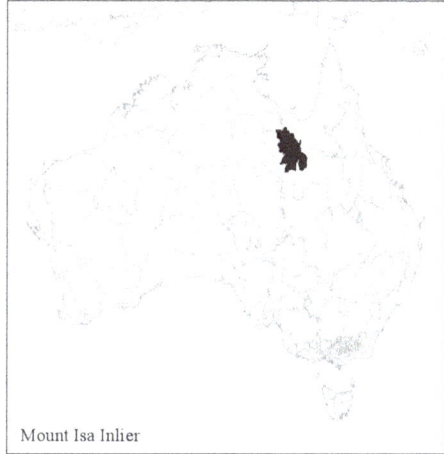

Mount Isa Inlier

Species 19   Specimens 67   Adults 67   Larvae 0

	J	F	M	A	M	J	J	A	S	O	N	D
Family Platycnemidae												
Nososticta fraterna				15								
Family Coenagrionidae												
Agriocnemis argentea				14								
Argiocnemis rubescens				1								
Austroagrion watsoni				2								
Austrocnemis maccullochi				1								
Ischnura aurora				2								
Ischnura heterosticta					1		1					
Pseudagrion aureofrons							6					
Pseudagrion jedda				4								
Pseudagrion microcephalum				7								
Family Aeshnidae												
Hemianax papuensis	1			1								
Family Gomphidae												
Austrogomphus mjobergi												1
Family Corduliidae												
Hemicordulia intermedia							1					

91

	J	F	M	A	M	J	J	A	S	O	N	D
Family Libellulidae												
Crocothemis nigrifrons				1		1						
Diplacodes haematodes				3								
Nannophlebia injibandi				1								
Orthetrum migratum				1								
Orthetrum sabina												
Orthetrum serapia												
Orthetrum villosovittatum												
Pantala flavescens												
Potamarcha congener												
Raphismia bispina												
Rhodothemis lieftincki				1								
Tramea loewii							1					

MGD   Mitchell Grass Downs   33,468,761 ha

Undulating downs on shales
and limestones; *Astrebla* spp.
grasslands and Acacia low
woodlands. Grey and brown
cracking clays.

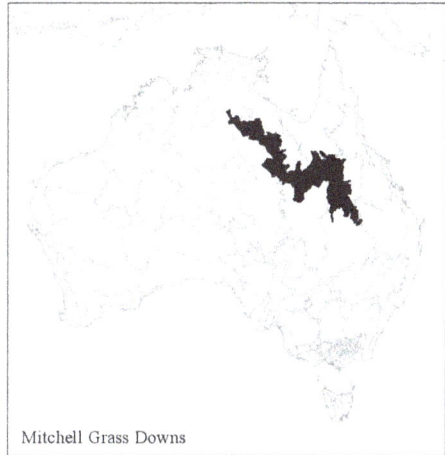

Mitchell Grass Downs

Species   14   Specimens   41   Adults   41   Larvae   0

	J	F	M	A	M	J	J	A	S	O	N	D	n.d.
Family Lestidae													
Austrolestes aridus		1			1			2		2			1
Lestes concinnus					1								
Family Coenagrionidae													
Ischnura aurora								1					
Ischnura heterosticta					1								
Xanthagrion erythroneurum				1	1					5			
Family Aeshnidae													
Austrogynacantha heterogena				1	1								
Hemianax papuensis				1	3		1						
Family Corduliidae													
Hemicordulia tau													1
Family Libellulidae													
Diplacodes bipunctata	1		1		1								
Diplacodes haematodes	2												
Orthetrum caledonicum				1	1					1			
Orthetrum sabina					1								
Pantala flavescens					3	1				1			1
Tholymis tillarga										1			

DEU   Desert Uplands   6,941,095 ha

Ranges and plains on dissected
Tertiary surface and Triassic
sandstones; woodlands of
Eucalyptus whitei, E.similis and
E.trachyphloia.

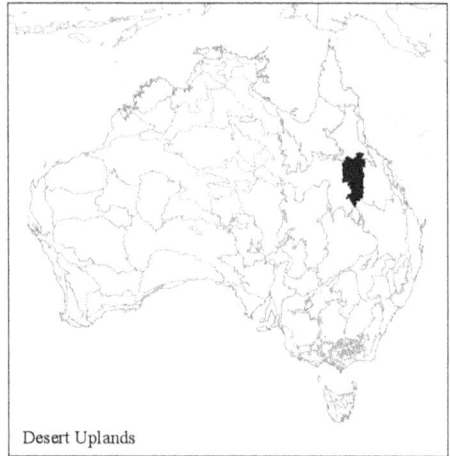

Desert Uplands

Species 32   Specimens 140   Adults 140   Larvae 0

	J	F	M	A	M	J	J	A	S	O	N	D
Family Lestidae												
Austrolestes aridus				2								
Austrolestes leda				3								
Lestes concinnus										1		
Family Argiolestidae												
Austroargiolestes icteromelas				5								
Family Isostictidae												
Austrosticta frater				1								
Eurysticta reevesi				1								
Rhadinosticta banksi				5								
Family Coenagrionidae												
Agriocnemis pygmaea				1								
Austroagrion watsoni				16								1
Ischnura aurora				7						1		
Pseudagrion ignifer				1								
Xanthagrion erythroneurum				3								
Family Aeshnidae												
Austrogynacantha heterogena				15						1		
Gynacantha nourlangie				1								

	J	F	M	A	M	J	J	A	S	O	N	D
Hemianax papuensis				3								
Family Gomphidae												
Ictinogomphus australis				4								
Family Synthemistidae												
Choristhemis flavoterminata				2								
Family Corduliidae												
Hemicordulia intermedia				3								
Hemicordulia tau				2								
Family Libellulidae												
Diplacodes bipunctata		1		1								
Diplacodes haematodes				11						1		3
Diplacodes trivialis				1								
Orthetrum caledonicum				7						1		2
Orthetrum villosovittatum				3								
Pantala flavescens		1		4								
Tramea loewii				10								1
Tramea stenoloba				1								
Zyxomma elgneri				8								

BBN   Brigalow Belt North   13,674,533 ha

Permian volcanics and Permian-Triassic sediments of the Bowen and Galilee Basins, Carboniferous and Devonian sediments and volcanics of the Drummond Basin and coastal blocks, Cambrian and Ordovician rocks of the Anakie inlier and associated Tertiary deposits. Subhumid to semiarid. Woodlands of ironbarks (*Eucalyptus melanophloia, E. crebra*), poplar box and Brown's box (*E. populnea, E. brownii*) and brigalow (*Acacia harpophylla*), blackwood (*A. argyrodendron*) and gidgee (*A. cambagei*). Region reaches the coast in the dry coastal corridor of Proserpine - Townsville.

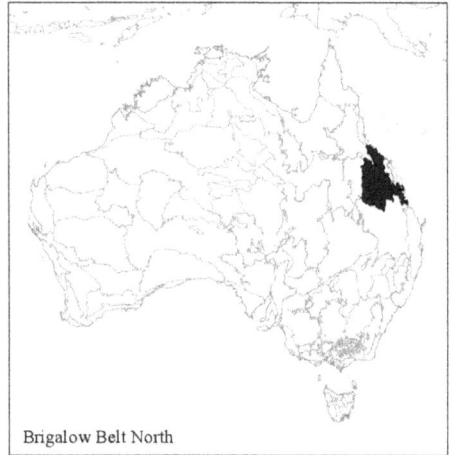

Brigalow Belt North

Species   90   Specimens 796   Adults   796   Larvae   0

	J	F	M	A	M	J	J	A	S	O	N	D	n.d.
Family Synlestidae													
Episynlestes albicauda											1	1	
Synlestes selysi	1	1											
Family Lestidae													
Austrolestes aridus							1						
Austrolestes insularis											1		
Austrolestes leda	1	3	4		1		2	1					
Lestes concinnus				1						2	2		
Family Argiolestidae													
Austroargiolestes icteromelas	2	1	1		1	1					4	1	
Family Lestoideidae													
Diphlebia coerulescens											1		
Family Isostictidae													
Labidiosticta vallisi	2	1	2	6					1				
Rhadinosticta banksi	2		3	2									
Rhadinosticta simplex	1										1	1	

	J	F	M	A	M	J	J	A	S	O	N	D	n.d.
Family Platycnemidae													
Nososticta solida	5		7	1	1				4	6	6	1	
Nososticta solitaria		1	1						1		2	1	
Family Coenagrionidae													
Aciagrion fragile										2			
Agriocnemis argentea					1				1				
Agriocnemis dobsoni			1										
Agriocnemis pygmaea	1		2		1	1	2		2		1	1	
Agriocnemis rubricauda			1										
Argiocnemis rubescens		1	1	1			1		1		1		
Austroagrion exclamationis	1										2		1
Austroagrion watsoni		1			1								
Austrocnemis splendida					1		2				2		1
Ceriagrion aeruginosum	2		9				4						
Ischnura aurora	2		2	5	1		1			3	3	2	
Ischnura heterosticta	3		2	2	2		3	5	1	3	6	2	1
Ischnura pruinescens			1	6			1						1
Pseudagrion aureofrons	1		1		1				1	1	4		
Pseudagrion ignifer		1	1		2					1	2		
Pseudagrion microcephalum					1		6					1	
Xanthagrion erythroneurum			1							1			
Family Aeshnidae													
Adversaeschna brevistyla		1											
Anax gibbosulus				2					2				
Anax guttatus				1									
Antipodophlebia asthenes											1		
Austrogynacantha heterogena	3	1	4	2					1	4	4	8	3
Austrophlebia costalis	1												
Gynacantha dobsoni		3		1					1		1		
Gynacantha rosenbergi										1			
Hemianax papuensis	2	1	2	1	1	1	1	1		1		4	2
Family Petaluridae													
Petalura ingentissima		1											
Family Gomphidae													
Antipodogomphus acolythus	1									1	1		
Antipodogomphus proselythus	1												
Austroepigomphus praeruptus								1		1			
Austroepigomphus turneri										5		2	

	J	F	M	A	M	J	J	A	S	O	N	D	n.d.
Austrogomphus amphiclitus			2	1							2		
Austrogomphus arbustorum											1		
Austrogomphus australis	2										1		
Austrogomphus cornutus											4	1	
Hemigomphus heteroclytus			1							2		1	
Ictinogomphus australis	5	3		3	1					2	2	3	4
Family Synthemistidae													
Choristhemis flavoterminata	3	1	1	1							2	1	
Eusynthemis nigra											4		7
Parasynthemis regina											1		
Family Macromiidae													
Macromia tillyardi											1		
Family Corduliidae													
Hemicordulia australiae		3	2		1					1		1	
Hemicordulia intermedia	1		3								4		1
Hemicordulia tau											1	1	
Family Libellulidae													
Aethriamanta circumsignata													1
Aethriamanta nymphaea										1		1	
Agrionoptera i.allogenes			1										
Brachydiplax denticauda	1		1	3	3		7				2		
Crocothemis nigrifrons	5	1	1	2	2		1	2	2	2	1	2	
Diplacodes bipunctata	7	3	4	5				2	7	2	4		
Diplacodes haematodes	6	8	5	6	2		1		1	10	5	1	
Diplacodes trivialis	6	2	3	6	5	2				1	1		
Hydrobasileus brevistylus		2	2	3					1	1		3	1
Lathrecista asiatica festa		1					1				1		
Macrodiplax cora	2	1									1		
Nannodiplax rubra		2					1	1		1			
Nannophlebia risi			1		1					2	4	1	
Neurothemis stigmatizans	4	4		1			1	4	2		3	1	
Orthetrum caledonicum	4	2	3	6		1			1	2	2	2	1
Orthetrum migratum										1			
Orthetrum sabina		4	1	3	1		2			1			1
Orthetrum serapia		1					1	1					
Orthetrum villosovittatum	2	2	1						3	3		2	
Pantala flavescens		1	2	2							2		3
Potamarcha congener		19								1	1		

	J	F	M	A	M	J	J	A	S	O	N	D	n.d.
Rhodothemis lieftincki	2	2	1	2						1	1	1	
Rhyothemis braganza	2									1	2	1	
Rhyothemis graphiptera	8	4	1	3	1			5	1	11	4	2	
Rhyothemis phyllis	3	1		3	1				1			1	1
Rhyothemis princeps	3			3									1
Tholymis tillarga		2						1				1	
Tramea loewii	3	3	3	5	3			1		1	2	2	1
Tramea stenoloba				1									
Urothemis aliena											1		
Zyxomma elgneri											1		
Genera incertae sedis													
Cordulephya pygmaea					1								
Micromidia·atrifrons	1	3	3	2							3		

99

CMC   Central Mackay Coast   1,464,208 ha

Humid tropical coastal ranges and plains. Rainforests (complex evergreen and semi-deciduous notophyll vine forest), *Eucalyptus* open forests and woodlands, *Melaleuca* spp. wetlands.

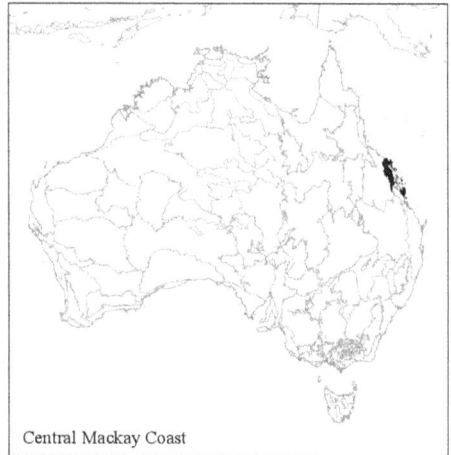

Central Mackay Coast

Species   71   Specimens   622   Adults   622   Larvae   0

	J	F	M	A	M	J	J	A	S	O	N	D	n.d.
Family Synlestidae													
Episynlestes intermedius	1									1	4	1	1
Synlestes selysi	3		1								1		
Family Lestidae													
Austrolestes leda	1												
Lestes concinnus											1		
Family Argiolestidae													
Austroargiolestes elke										5	6		1
Austroargiolestes icteromelas	6		11							9	11	3	
Family Lestoideidae													
Diphlebia coerulescens	11	2	11							4	7	4	
Family Isostictidae													
Labidiosticta vallisi	2				3					1	1		
Rhadinosticta simplex	3										1		
Family Platycnemidae													
Nososticta solida	4	2	3							3	1	2	
Nososticta solitaria	4		3							2	6	22	
Family Coenagrionidae													
Agriocnemis argentea	2	1								3		7	

	J	F	M	A	M	J	J	A	S	O	N	D	n.d.
Agriocnemis pygmaea	2												
Argiocnemis rubescens	3		2			1				1			
Austroagrion watsoni	2									3	1		
Austrocnemis splendida					7								
Ischnura aurora		1								2			
Ischnura heterosticta	2		5		1	1	1	1		1	1		
Pseudagrion aureofrons		1								3			
Pseudagrion ignifer	2		5		2					3	1		
Pseudagrion microcephalum	2									5	1	2	
Xanthagrion erythroneurum											1		
Family Aeshnidae													
Adversaeschna brevistyla	1		1										
Austroaeschna christine	3		1	1							2		
Austroaeschna eungella	2			1						1	1	3	
Austroaeschna pinheyi	2												
Austrophlebia costalis	1									1	3	2	1
Hemianax papuensis	2		1							1			
Telephlebia cyclops				1						1		2	
Family Petaluridae													
Petalura litorea			1										1
Family Gomphidae													
Austroepigomphus turneri		1									3	1	
Austrogomphus amphiclitus	2		1							1	3	4	1
Hemigomphus comitatus										1			
Hemigomphus heteroclytus	6									5	2	2	
Ictinogomphus australis		1								1	2		
Family Synthemistidae													
Choristhemis flavoterminata	4		1							2	4	3	
Eusynthemis nigra	12		2							1	17	7	12
Family Corduliidae													
Hemicordulia australiae	3		1										
Hemicordulia intermedia											1		
Hemicordulia tau											1		
Family Libellulidae													
Brachydiplax denticauda			1							1	5	1	
Crocothemis nigrifrons	2			1	1					4	1	3	
Diplacodes bipunctata	2		1	1	1			12		3	5		
Diplacodes haematodes	5		1		1					4	3	1	

	J	F	M	A	M	J	J	A	S	O	N	D	n.d.
Diplacodes trivialis			1					8		4	11		
Hydrobasileus brevistylus										1	4	1	
Lathrecista asiatica festa										6			
Nannodiplax rubra			2			1				1			
Nannophlebia eludens	2		1								2	2	
Nannophlebia risi	1	1								5	1	1	1
Nannophya australis					4								
Neurothemis stigmatizans					1		2			6			1
Orthetrum caledonicum	3	1			1			3		2	2		
Orthetrum migratum			2										
Orthetrum sabina	2		1	1						5	1		1
Orthetrum serapia										1	1		
Orthetrum villosovittatum	3		4							1	4	2	
Pantala flavescens			1	2						1	1		
Rhodothemis lieftincki										1	1		
Rhyothemis braganza							1						
Rhyothemis graphiptera	2		1							1	5		1
Rhyothemis phyllis			2	1						1	4	4	
Rhyothemis princeps											2		1
Rhyothemis resplendens												3	
Tholymis tillarga			1								2		1
Tramea eurybia		1											
Tramea loewii				3	1	1					1	1	1
Genera incertae sedis													
Austrocordulia refracta		1								1			

CHC   Channel Country   30,409,437 ha

Low hills on Cretaceous sediments; forbfields and Mitchell grass downs, and intervening braided river systems of coolibah *E.coolibah* woodlands and lignum/saltbush *Muehlenbeckia* sp./ *Chenopodium* sp. shrublands. (Includes small areas of sand plains.)

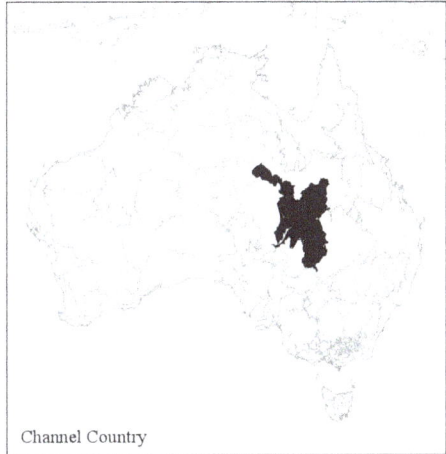

Channel Country

Species   15   Specimens   65   Adults   65   Larvae   0

	J	F	M	A	M	J	J	A	S	O	N	D	n.d.
Family Lestidae													
Austrolestes annulosus										1			
Austrolestes aridus									1	3	1		
Austrolestes leda									1				
Family Coenagrionidae													
Ischnura aurora			1						2	1	2		
Ischnura heterosticta									1				
Pseudagrion aureofrons										1			
Xanthagrion erythroneurum	3	1								4			
Family Aeshnidae													
Adversaeschna brevistyla									1				
Hemianax papuensis			1		2			1			2		1
Family Gomphidae													
Austrogomphus australis										3	2		1
Family Corduliidae													
Hemicordulia tau	1		1	4	2					2	1		
Family Libellulidae													
Diplacodes bipunctata			2		3			2	1		1		

	J	F	M	A	M	J	J	A	S	O	N	D	n.d.
Diplacodes haematodes					1				2				
Orthetrum caledonicum			1	1				1		1			

MUL   Mulga Lands   25,188,333 ha

Undulating plains and low hills on Cainozoic sediments; red earths and lithosols; *Acacia aneura* shrublands and low woodlands.

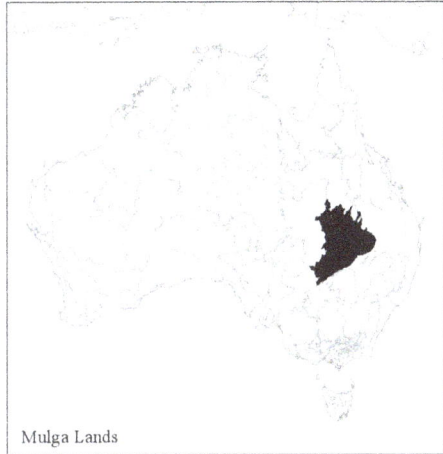

Mulga Lands

Species   25   Specimens   514   Adults   257   Larvae   257

	J	F	M	A	M	J	J	A	S	O	N	D	n.d.
Family Lestidae													
Austrolestes annulosus									1				
Austrolestes aridus				1					2	5			
Austrolestes leda	1				1				2	1	3		
Family Isostictidae													
Rhadinosticta simplex										2			
Family Platycnemidae													
Nososticta solida										2			
Family Coenagrionidae													
Austroagrion watsoni										1			
Ischnura aurora			2		29				13	2	6	2	
Ischnura heterosticta									2		1		
Pseudagrion aureofrons										6			
Pseudagrion microcephalum					2								
Xanthagrion erythroneurum		1			3				16		1		1
Family Aeshnidae													
Austrogynacantha heterogena	1											1	
Hemianax papuensis		1		3	6				4		1		2
Family Gomphidae													
Austrogomphus australis										9			

	J	F	M	A	M	J	J	A	S	O	N	D	n.d.
Family Synthemistidae													
Eusynthemis nigra													1
Family Corduliidae													
Hemicordulia tau					3				10		1		
Family Libellulidae													
Diplacodes bipunctata			2		32				11		7		1
Diplacodes haematodes					2				1			4	
Nannophya fenshami										7			
Orthetrum caledonicum				2	13				6	1	1	4	
Orthetrum migratum					1								
Pantala flavescens					2				2				
Tramea loewii									5	1			

BBS   Brigalow Belt South   27,219,776 ha

Predominantly Jurassic and younger deposits of the Great Artesian Basin and Tertiary deposits with elevated basalt flows. Subhumid. Eucalyptus woodlands and open forests of ironbarks, poplar box, spotted gum (*E. maculata*), cypress pine (*Callitris glaucophylla*), Bloodwoods (eg. *E. trachyphloia*, *E. hendersonii* ms) brigalow-belah forests (*E. harpophylla*, *Casuarina cristata*) and semi-evergreen vine thicket.

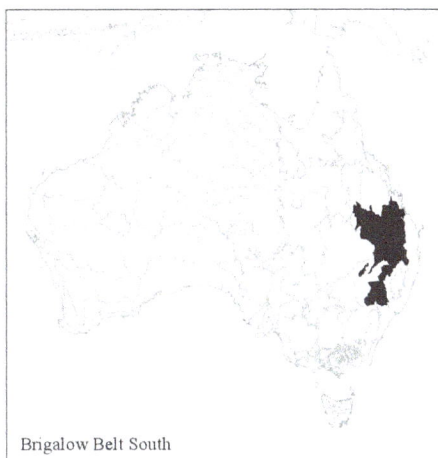

Brigalow Belt South

Species   94   Specimens   2752   Adults   2206   Larvae   546

	J	F	M	A	M	J	J	A	S	O	N	D	n.d.
Family Synlestidae													
Episynlestes albicauda	1	2	7	4	1					3	1	6	
Synlestes selysi	1			2	1	2							
Synlestes weyersii	1	5	23	1					1		5	3	5
Family Lestidae													
Austrolestes analis			1									2	
Austrolestes aridus	3			2	1	1			3	2	1	3	2
Austrolestes leda	9	3	8	11	7	2	2	1	5	10	39	24	2
Austrolestes psyche	4										1		
Lestes concinnus						1			2			1	
Family Argiolestidae													
Austroargiolestes amabilis	1		1										
Austroargiolestes icteromelas	14	11	35	22	1				10	39	15	13	9
Griseargiolestes eboracus											1		
Griseargiolestes fontanus	1	1											
Family Lestoideidae													
Diphlebia coerulescens												1	
Diphlebia lestoides	1										1	4	4

	J	F	M	A	M	J	J	A	S	O	N	D	n.d.
Diphlebia nymphoides	1	4	5	1					2	17		7	
Family Isostictidae													
Labidiosticta vallisi										1			
Rhadinosticta simplex	1									1	3		
Family Platycnemidae													
Nososticta solida	9	6	15	3					2	13	7	4	
Family Coenagrionidae													
Agriocnemis argentea										1			
Agriocnemis pygmaea			4	3							6		
Agriocnemis rubricauda			2										
Argiocnemis rubescens			1	1								2	
Austroagrion watsoni	4	1	1	7	1				6	6	11	7	1
Austrocnemis splendida									1				
Ceriagrion aeruginosum													1
Ischnura aurora	1	1	8	15	2	1			6	6	33	21	1
Ischnura heterosticta	8	9	13	9	5			1	5	14	25	5	3
Ischnura pruinescens	1												
Pseudagrion aureofrons	12	9	11	5						3	4	9	
Pseudagrion cingillum	2											7	
Pseudagrion ignifer		3		2					1		2	5	1
Pseudagrion microcephalum				2					1		1		
Xanthagrion erythroneurum	3	1	7	4	1				3	2	22	10	
Family Aeshnidae													
Adversaeschna brevistyla	6	1		9							6	2	3
Austroaeschna parvistigma												1	
Austroaeschna pinheyi	1	47	4	4						10	2		
Austroaeschna sigma		2		1									
Austroaeschna unicornis		4	1	3									1
Austroaeschna muelleri		2	6	2					1	3	1	6	3
Austroaeschna pulchra	2	1	5	1								1	1
Austrogynacantha heterogena	4	2	13								2	5	3
Dendroaeschna conspersa			3										
Gynacantha dobsoni		1											
Hemianax papuensis	6	6	14	9	1			1	4	7	2	10	4
Spinaeschna tripunctata													1
Telephlebia cyclops	2	3											
Telephlebia undia												1	
Family Gomphidae													

	J	F	M	A	M	J	J	A	S	O	N	D	n.d.
Antipodogomphus acolythus		2	1										
Antipodogomphus proselythus										1			
Austroepigomphus praeruptus	7	2	1	1							2	7	1
Austrogomphus amphiclitus		2	8	5						3	2	7	1
Austrogomphus arbustorum	2												
Austrogomphus australis	1		2								4		
Austrogomphus cornutus	1	2	8							18	5	8	4
Austrogomphus guerini	2	4											
Austrogomphus ochraceus	1	5									2	1	3
Hemigomphus gouldii	2	2	1								1	3	
Hemigomphus heteroclytus	7	3	10	6					1	5	3	13	2
Ictinogomphus australis	4	1	3	3							1	9	2
Family Synthemistidae													
Choristhemis flavoterminata	1	3	5	2							4	9	3
Eusynthemis aurolineata	6											1	1
Eusynthemis deniseae		1	1							5	1	9	2
Eusynthemis nigra	3	6	1							1	3	26	11
Eusynthemis virgula	1												3
Parasynthemis regina	4		1	1	2					3	1	1	
Family Macromiidae													
Macromia tillyardi	1										1	2	
Family Corduliidae													
Hemicordulia australiae	6	1	1	2						3	4	1	2
Hemicordulia intermedia	1	2	4	3						2	2	3	
Hemicordulia tau	5	5	6	11	6	1				9	10	16	4
Family Libellulidae													
Brachydiplax denticauda			1										
Crocothemis nigrifrons	2		7	4		1	1		2	5	2	1	
Diplacodes bipunctata	8	8	13	18	9	7	1		9	4	30	20	7
Diplacodes haematodes	13	4	15	25	1	3			11	30	11	7	
Diplacodes melanopsis	1		3	1								1	
Hydrobasileus brevistylus	1												
Nannodiplax rubra											1		
Nannophlebia risi		1								3	1	3	1
Nannophya australis	1										9		1
Orthetrum caledonicum	8	3	18	34					7	17	27	12	4
Orthetrum migratum	6										1		
Orthetrum sabina									1		1	1	

	J	F	M	A	M	J	J	A	S	O	N	D	n.d.
Orthetrum villosovittatum	3		6	26					1	5	16	2	2
Pantala flavescens	2	1	5	1							1		1
Potamarcha congener		1											
Rhodothemis lieftincki	1	1	1									3	1
Rhyothemis graphiptera	4		4								1		
Rhyothemis phyllis											1		
Rhyothemis princeps								1					
Tholymis tillarga			3										
Tramea loewii	1	2	8	3					8	7	5	3	2
Zyxomma elgneri	2	2	7	4						1	1	4	
Genera incertae sedis													
Austrocordulia refracta			1							1		1	
Cordulephya pygmaea		1											

SEQ   South Eastern Queensland   7,804,921 ha

Metamorphic and acid to
basic volcanic hills and ranges
(Beenleigh, D'Aguilar, Gympie,
Yarraman Blocks) sediments
of the Moreton, Nambour and
Maryborough Basins, extensive
alluvial valleys and Quaternary
coastal deposits including high
dunes on the sand islands
such as Fraser Island. Humid.
*Eucalyptus-Lophostemon-
Syncarpia* tall open forests,
Eucalyptus open forests
and woodlands, sub-tropical
rainforests often with *Araucaria
cunninghamii* emergents
and small areas of cool temperate rainforest dominated by
Nothofagus moorei and semi-evergreen vine thickets, *Melaleuca
quinquenervia* wetlands and Banksia low woodlands, heaths and
mangrove/saltmarsh communities.

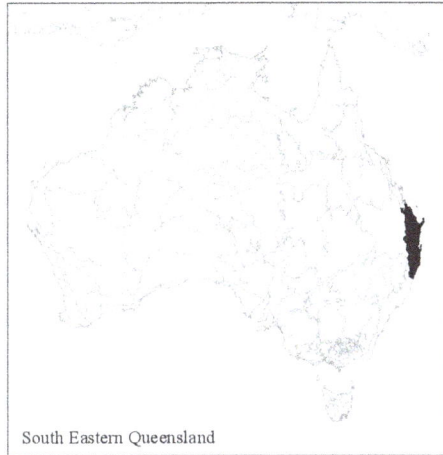

South Eastern Queensland

Species   130   Specimens   8957   Adults   8930   Larvae   27

	J	F	M	A	M	J	J	A	S	O	N	D	n.d.
Family Synlestidae													
Episynlestes albicauda	17	10	15	30	21	7	2	3	1	8	3	28	5
Synlestes selysi	18	3	17	17	31	5				3	2	20	3
Synlestes weyersii	5	11	14	12	1				1		3	1	1
Family Lestidae													
Austrolestes aridus	1			2								1	
Austrolestes cingulatus								1					
Austrolestes leda	32	10	25	32	37	3	7	9	41	13	19	46	7
Austrolestes minjerriba	54	1	15	16	13			2	37	6	7	40	1
Austrolestes psyche	8	1		4					1	4	1	9	1
Lestes concinnus				1						2		1	
Family Argiolestidae													
Austroargiolestes amabilis	14	6		1		1				10	14	18	1
Austroargiolestes chrysoides	11	4	14	1		1			28	13	15	19	
Austroargiolestes icteromelas	89	39	55	63	27	1	1	2	33	46	76	93	2

	J	F	M	A	M	J	J	A	S	O	N	D	n.d.
Griseargiolestes albescens	22	7	5	3	6			2	35	18	4	24	2
Griseargiolestes fontanus	14	1		1						1	3	13	
Family Lestoideidae													
Diphlebia coerulescens	32	7	22	5	4		3			9	21	29	
Diphlebia lestoides	14	1							1	4	20	20	2
Diphlebia nymphoides	12	7	2	1					6	1	6	7	
Family Isostictidae													
Labidiosticta vallisi	4	3	1	11	8					2	1	1	1
Neosticta canescens	16	1								5	26	11	2
Rhadinosticta simplex	14	6	4	2	8				3		2	9	
Family Platycnemidae													
Nososticta coelestina			6										
Nososticta solida	52	10	7	10		1		1	1	15	28	52	1
Nososticta solitaria	8		7						2			3	2
Family Coenagrionidae													
Aciagrion fragile	6						1		3	2			
Agriocnemis argentea			1							1			
Agriocnemis pygmaea	27	11	5	28					8	9	14	26	
Agriocnemis rubricauda	1								1	1	2	1	
Argiocnemis rubescens	16	9	9	18	2	1	1		13	10	15	22	1
Austroagrion exclamationis	12		3	1						1	1	12	
Austroagrion watsoni	23	6	6	4	5	2			32	36	35	33	2
Austrocnemis splendida	10	10	4	2					4	18	7	15	
Ausrocoenagrion lyelli													1
Caliagrion billinghursti										1			
Ceriagrion aeruginosum	21	7	12	6	1	1				5	12	14	
Ischnura aurora	20	9	5	5	5	3		2	11	10	13	27	1
Ischnura heterosticta	56	21	28	31	29	7		3	33	27	40	39	3
Ischnura pruinescens	9	1		1		1							
Pseudagrion aureofrons	15	13	3	8	1	2				8	6	23	3
Pseudagrion cingillum	1			1		2			1		1	2	
Pseudagrion ignifer	34	10	18	39	7	1				5	16	32	
Pseudagrion microcephalum	12	15	14	16	3	4		3	5	7	19	39	4
Xanthagrion erythroneurum	3	2	2	4	1			1	1	2	4	4	2
Family Aeshnidae													
Adversaeschna brevistyla	22	8	9	6	3	2		3	8	2	10	18	2
Anax gibbosulus		1	1	1									
Anax guttatus	1												

	J	F	M	A	M	J	J	A	S	O	N	D	n.d.
Antipodophlebia asthenes	5	1									3	7	
Austroaeschna cooloola	3	1								1	2	4	2
Austroaeschna pinheyi	4	5	9	1	9	1				1		3	
Austroaeschna pulchra	10	5	19	22	7	1				1		7	1
Austroaeschna sigma	12	4	8	13	12	3		3			2	13	
Austroaeschna unicornis	4		3										
Austrogynacantha heterogena	12	10	2	13	2						2	3	1
Austrophlebia costalis	7	2						1		2	1	9	2
Dendroaeschna conspersa			3	2	3								
Gynacantha dobsoni												2	
Hemianax papuensis	12	9	28	18	8	3	3	2	3	5	2	11	2
Notoaeschna geminata	1							1	1			2	
Spinaeschna tripunctata		1											
Telephlebia cyclops	18	13	5	6	1	1			1			4	12
Telephlebia tryoni	4	4									6	9	6
Family Petaluridae													
Petalura gigantea	2											3	
Petalura litorea	15	1								2	1	10	1
Family Gomphidae													
Antipodogomphus acolythus	3		4	1								3	
Antipodogomphus proselythus	2	1								1	2		1
Austroepigomphus praeruptus	8	1	3	2					1	7	3	11	1
Austrogomphus amphiclitus	62	21	13	15	8				3	10	14	40	2
Austrogomphus australis	2												
Austrogomphus cornutus	11		5	2						1	6	10	
Austrogomphus guerini	1		3										
Austrogomphus ochraceus	19	8	2					1	12	13	20	48	1
Hemigomphus cooloola	5	1									1	6	
Hemigomphus gouldii	5	2	2					1		4	6	5	
Hemigomphus heteroclytus	16	8	4	10	1				4	4	15	18	3
Ictinogomphus australis	29	11	10	5	1						1	17	1
Family Synthemistidae													
Choristhemis flavoterminata	56	16	13	8	3		1			5	31	58	2
Eusynthemis aurolineata	4	1	4	5						1	8	13	
Eusynthemis brevistyla	1												
Eusynthemis cooloola												1	
Eusynthemis nigra	49	1	5	5				2		5	35	55	9
Eusynthemis rentziana	2								1			1	

	J	F	M	A	M	J	J	A	S	O	N	D	n.d.
Eusynthemis virgula	1											1	
Parasynthemis regina	3	5					1				6	4	
Tonyosynthemis ofarrelli	3											1	1
Family Macromiidae													
Macromia tillyardi		1											
Family Corduliidae													
Hemicordulia australiae	55	14	17	24	12	1		1	22	27	46	45	3
Hemicordulia continentalis	26	11	5	2	2				2	2	6	20	
Hemicordulia intermedia			1		1								
Hemicordulia superba	6	2	1	3							2		3
Hemicordulia tau			3	2	3				3		1	5	
Metaphya tillyardi													
Procordulia jacksoniensis											1		
Family Libellulidae													
Aethriamanta circumsignata	1	1								2	1	2	
Aethriamanta nymphaea	3	4	5		1					1	1	3	
Agrionoptera i.allogenes	4	12	2	2	6					1		1	
Austrothemis nigrescens	1	1			1			7	1	2		13	2
Brachydiplax denticauda	10	15	12	1	2					2	6	11	
Crocothemis nigrifrons	16	18	15	25	5	3	1		1	6	13	10	
Diplacodes bipunctata	49	23	72	68	36	2	3	16	40	17	28	32	4
Diplacodes haematodes	38	10	48	59	26	3	2	8	9	24	22	45	3
Diplacodes melanopsis	10	13	5	2				1	1	25	12	18	7
Diplacodes trivialis	2	8	3	4	4				2		1		
Hydrobasileus brevistylus	14	9	5	1			2				2	10	5
Lathrecista asiatica festa					1								
Macrodiplax cora	3	3	2	3	2	1					2		
Nannodiplax rubra	31	4	16	3				1	28		8	35	1
Nannophlebia risi	30	7	11	15	11				1	4	4	15	2
Nannophya australis	40	2	5	4	2			1	35	16	2	50	
Orthetrum boumiera	43	8	31	5	1					7	5	44	1
Orthetrum caledonicum	5	38	35	44	14	2	1	2	11	13	26	46	3
Orthetrum migratum												1	
Orthetrum sabina	24	11	26	15	2			1	4	5	9	11	
Orthetrum serapia			2										
Orthetrum villosovittatum	58	41	42	55	17	2	1	3	9	11	32	56	5
Pantala flavescens	5	4	10	6	2			2	1		1		
Potamarcha congener					1								

	J	F	M	A	M	J	J	A	S	O	N	D	n.d.
Rhodothemis lieftincki	14	13	9	4				1		3	1	8	
Rhyothemis graphiptera	41	5	14	6	3			2	1	1	17	27	4
Rhyothemis Phyllis chloe	37	15	5	1						1	11	30	3
Rhyothemis princeps							1						
Tholymis tillarga					1								
Tramea eurybia	43	1	1	2				1			5	32	1
Tramea loewii	35	15	29	14	6			2	18	21	9	20	2
Tramea stenoloba	6			1									
Urothemis aliena	3			1							3	2	1
Zyxomma elgneri	26	10	4	1						1	1	5	
Genera incertae sedis													
Austrocordulia refracta	10	1								1	3	3	3
Cordulephya pygmaea	1		5	10	12								6
Micromidia·atrifrons	4	8	3	2							1	3	1
Micromidia convergens	7	3										2	

DRP   Darling Riverine Plains 10,699,769 ha

Alluvial fans and plains;
summer/winter rainfall
in catchments, including
occasional cyclonic influence;
grey clays; woodlands and
open woodlands dominated by
Eucalyptus spp.

Darling Riverine Plains

Species   28   Specimens   555   Adults   70   Larvae   485

	J	F	M	A	M	J	J	A	S	O	N	D	n.d.
Family Lestidae													
Austrolestes analis									1				
Austrolestes aridus			2						1				
Austrolestes leda						1			3	2	1		
Austrolestes psyche	2												
Family Platycnemidae													
Nososticta solida	1												
Family Coenagrionidae													
Ischnura aurora	1	1	2		1						1	1	
Ischnura heterosticta		1										1	
Xanthagrion erythroneurum	4	1						1				4	
Family Aeshnidae													
Austrogynacantha heterogena	2												
Hemianax papuensis	1		2						1				1
Family Gomphidae													
Antipodogomphus acolythus										1			
Austrogomphus australis										1		2	
Family Synthemistidae													
Parasynthemis regina													1

	J	F	M	A	M	J	J	A	S	O	N	D	n.d.
Family Corduliidae													
Hemicordulia tau	1			1					1	1			
Family Libellulidae													
Diplacodes bipunctata	1		2		3			1				2	
Diplacodes haematodes	1												
Orthetrum caledonicum	2		1		2			1				1	

NAN   Nandewar   2,701,977 ha

Hills on Palaeozoic sediments;
lithosols and earths;
Eucalyptus albens woodlands;
summer rainfall.

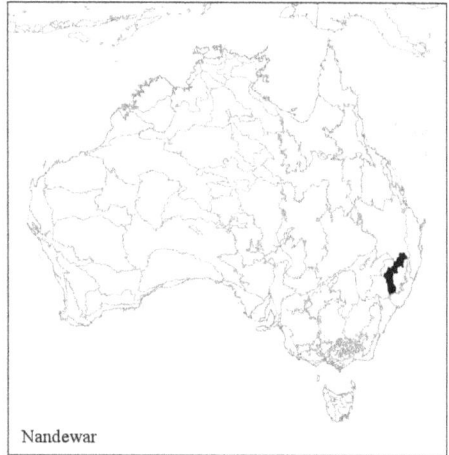

Nandewar

Species 61   Specimens 942   Adults 549   Larvae 393

	J	F	M	A	M	J	J	A	S	O	N	D	n.d.
Order Odonata													
Family Synlestidae													
Episynlestes albicauda											1		
Synlestes selysi												1	
Synlestes weyersii	1	3	1										
Family Lestidae													
Austrolestes analis	1	1											
Austrolestes annulosus	1												
Austrolestes cingulatus	2	1											
Austrolestes leda	1	2	1								2		1
Austrolestes psyche	7			1							2		
Family Argiolestidae													
Austroargiolestes brookhousei	1												
Austroargiolestes icteromelas	10	3	2	1					1	3	5	1	
Griseargiolestes eboracus		3											
Family Lestoideidae													
Diphlebia lestoides			1										
Diphlebia nymphoides	5		1	1								8	

	J	F	M	A	M	J	J	A	S	O	N	D	n.d.
Family Platycnemidae													
Nososticta solida	16	4	4	4	1						10	2	
Family Coenagrionidae													
Austroagrion watsoni	1		2	13	2						6		
Ischnura aurora				2									
Ischnura heterosticta			5	4									
Pseudagrion aureofrons	6		4	4	1						7	20	
Pseudagrion cingillum													
Xanthagrion erythroneurum	1		6		1					4	9		
Family Aeshnidae													
Austroaeschna parvistigma		1				1							
Austroaeschna pulchra		2	1										
Austroaeschna sigma	2		4	1									3
Austroaeschna subapicalis		2											
Austroaeschna unicornis	1		1										
Hemianax papuensis	4	4	3		2					2	5		
Notoaeschna geminata	1												
Telephlebia godeffroyi		3					1						
Family Gomphidae													
Antipodogomphus acolythus											5		
Antipodogomphus dentosus													
Austroepigomphus praeruptus	1										1		
Austrogomphus amphiclitus	1	1	1								11		
Austrogomphus australis	2		1								3	2	
Austrogomphus cornutus	2	2	10								5	14	
Austrogomphus divaricatus													
Austrogomphus guerini											1		1
Austrogomphus ochraceus	1	11											
Hemigomphus gouldii											1		
Hemigomphus heteroclytus	10	4	4	2	2						8	6	
Family Synthemistidae													
Eusynthemis aurolineata	4	2										1	1
Eusynthemis brevistyla		1											
Parasynthemis regina				1								2	
Synthemis eustalacta		2											
Family Corduliidae													
Hemicordulia australiae				1	1								
Hemicordulia intermedia												4	

	J	F	M	A	M	J	J	A	S	O	N	D	n.d.
Hemicordulia superba			1								1	4	
Hemicordulia tau	1	1		1	3				1	1	3		1
Family Libellulidae													
Crocothemis nigrifrons		1	1									2	
Diplacodes bipunctata	1	4	5	1	3				2	5	8		
Diplacodes haematodes	14	4	5	5	2				3	4	18		
Diplacodes melanopsis	1												
Nannophlebia risi		1	1								2		
Orthetrum caledonicum	10	3	4	5	2					3	9		1
Orthetrum villosovittatum	1		1	1							1		
Pantala flavescens	1												
Tramea loewii			1										
Genera incertae sedis													
Apocordulia macrops													1

NET   New England Tablelands   3,002,213 ha

Elevated plateau of hills
and plains on Palaeozoic
sediments, granites and
basalts; dominated by stringy
bark/peppermint/box species,
including *Eucalyptus caliginosa,
E. nova-anglica, E. melliodora*
and *E. blakleyi.*

New England Tablelands

Species   82   Specimens   2537   Adults   2318   Larvae   219

	J	F	M	A	M	J	J	A	S	O	N	D	n.d.
Family Synlestidae													
Episynlestes albicauda	3												
Synlestes selysi	3			1								1	
Synlestes weyersii	32	13	2	41							9	24	2
Family Lestidae													
Austrolestes analis	1	1	2							1	3		
Austrolestes annulosus		4		5	3					2		6	
Austrolestes cingulatus	36	17	4	4						9	18	13	3
Austrolestes leda	18	3	6	3				2	7	9	14		
Austrolestes minjerriba		1	1										
Austrolestes psyche	49	3		3									
Family Argiolestidae													
Austroargiolestes alpinus	13	2								1	1	1	1
Austroargiolestes amabilis	1										1	4	
Austroargiolestes brookhousei	2											1	
Austroargiolestes icteromelas	127	43	5	4					2	13	49	31	6
Griseargiolestes eboracus	42	3	1	1						5	8	21	2
Griseargiolestes fontanus	1											2	

	J	F	M	A	M	J	J	A	S	O	N	D	n.d.
Family Lestoideidae													
Diphlebia coerulescens												1	
Diphlebia lestoides	1									2	8	7	
Diphlebia nymphoides	18	3	1							2	12	6	
Family Isostictidae													
Neosticta canescens											5	8	
Rhadinosticta simplex	2	3											
Family Platycnemidae													
Nososticta solida	12	4	2	1		1					13	2	
Family Coenagrionidae													
Agriocnemis pygmaea											1		
Austroagrion watsoni	19	14	1	1					9	2	12	6	1
Ausrocoenagrion lyelli	2									1	3	2	2
Caliagrion billinghursti	3	4									1	2	1
Ischnura aurora	5	3	1	2					1	1	5	2	
Ischnura heterosticta	31	8		3						5	16	7	
Pseudagrion aureofrons	2	1	1	1							3	1	
Pseudagrion ignifer	1	3											
Pseudagrion microcephalum													1
Xanthagrion erythroneurum	2	9	1	3	1					1	10	8	
Family Aeshnidae													
Adversaeschna brevistyla	15	4	1	4						2	2	6	4
Austroaeschna parvistigma	9	10	3									5	2
Austroaeschna pulchra	11	4	1	2								7	2
Austroaeschna sigma	5	1										6	1
Austroaeschna subapicalis	3											1	
Austroaeschna unicornis	2	6	4									5	
Austrogynacantha heterogena	1						1						
Austrophlebia costalis	1												
Dendroaeschna conspersa			1										
Hemianax papuensis	2	5	4	1					3	2	5	4	1
Notoaeschna geminata	5	2								4	6	5	
Spinaeschna tripunctata		2									4	1	
Telephlebia godeffroyi	4		1							1	3		1
Family Petaluridae													
Petalura gigantea	5											21	2
Family Gomphidae													
Austroepigomphus praeruptus	7	2		1								1	
Austrogomphus amphiclitus	24	2	2	2								1	

	J	F	M	A	M	J	J	A	S	O	N	D	n.d.
Austrogomphus australis	1										4	5	1
Austrogomphus cornutus	4	2	9								1	6	
Austrogomphus guerini	57	42	1	1						2	9	18	3
Austrogomphus ochraceus	16	33	5							1	6	13	
Hemigomphus gouldii	12	2								2	6		1
Hemigomphus heteroclytus	12	8	2	2					1	1	4	3	1
Family Synthemistidae													
Archaeosynthemis orientalis	21	1								1	2	3	
Choristhemis flavoterminata	22	7	1										
Eusynthemis aurolineata	19	1									4	7	2
Eusynthemis brevistyla	1	1									1	7	
Eusynthemis nigra	1											1	
Eusynthemis rentziana												3	
Eusynthemis virgula	1	2									5	7	
Parasynthemis regina	6											2	
Synthemis eustalacta	36	8	4	2						1	5	16	5
Tonyosynthemis ofarrelli												2	
Family Corduliidae													
Hemicordulia australiae	18	7		1							4	1	2
Hemicordulia intermedia	1	2	1									4	
Hemicordulia tau	1	6	3	2					1	5	9	8	1
Procordulia jacksoniensis	11	3	1							1	3	1	
Family Libellulidae													
Diplacodes bipunctata	16	13	10	8	1					3	22	7	
Diplacodes haematodes	20	11	8	6					1	3	22	5	
Diplacodes melanopsis	4	1	3	2						2			
Diplacodes nebulosa													
Nannophlebia risi		2										1	
Nannophya australis	25	2	1							3	3		
Nannophya dalei	16	1								2	6	4	1
Orthetrum caledonicum	19	10	1	4					1	1	8	6	
Orthetrum villosovittatum	11	3		3							1	5	
Rhyothemis graphiptera	2												
Tramea loewii		1		2									
Genera incertae sedis													
Austrocordulia refracta	1									1			
Cordulephya montana	10											1	
Cordulephya pygmaea		2	4	1		1							
Micromidia·atrifrons													1

NNC NSW North Coast 3,996,591 ha

Humid; hills, coastal plains
and sand dunes; *Eucalyptus -
Lophostemon confertus* tall open
forests, *Eucalyptus* open forests
and woodlands, sub-tropical
rainforest often with *Araucaria
cunninghamii* (complex notophyll
and microphyll vine forest),
Melaleuca quinquenervia
wetlands, and heaths.

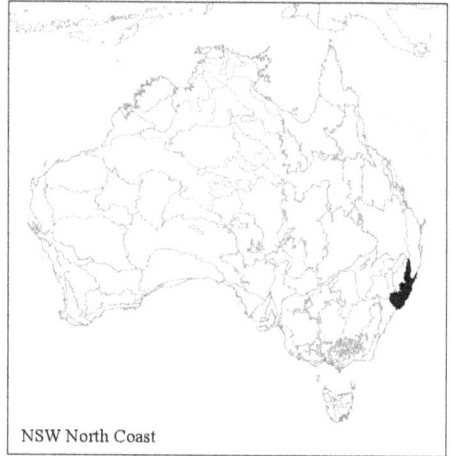

NSW North Coast

Species 108 Specimens 2805 Adults 2558 Larvae 247

	J	F	M	A	M	J	J	A	S	O	N	D	n.d.
Family Synlestidae													
Episynlestes albicauda	4		4		1							1	1
Synlestes selysi	5	1		2	1	4						1	
Synlestes weyersii	14	7	6	7	1						22	17	3
Family Lestidae													
Austrolestes analis	6	4									1	1	1
Austrolestes aridus											1		
Austrolestes cingulatus	2	3		1						2	16	5	2
Austrolestes leda	6	8		1						2	4	5	
Austrolestes minjerriba	1	2	1	1	1				2	4	1	1	
Austrolestes psyche	6	9		4	3	1				15	4	5	2
Family Argiolestidae													
Austroargiolestes alpinus		3											
Austroargiolestes amabilis	4	1									4	16	1
Austroargiolestes brookhousei	12										2	9	5
Austroargiolestes christine	3	1									1	17	27
Austroargiolestes icteromelas	26	15	9	2	1			1		20	51	137	9
Griseargiolestes albescens	2			2	2	1				4		1	
Griseargiolestes bucki	6	1									2	23	10

	J	F	M	A	M	J	J	A	S	O	N	D	n.d.
Griseargiolestes eboracus	23	5								2	6	9	15
Griseargiolestes fontanus												5	
Family Lestoideidae													
Diphlebia coerulescens	1									2		5	
Diphlebia lestoides	7	1								3	3	30	8
Diphlebia nymphoides	14	1								4	62	3	1
Family Isostictidae													
Neosticta canescens	5	1									10	9	1
Rhadinosticta simplex	13	15	1	1						3			
Family Platycnemidae													
Nososticta solida	12	7	1							1	41	7	
Family Coenagrionidae													
Agriocnemis pygmaea	5	11	2	3	2					3	2	2	
Argiocnemis rubescens	2	1	1		1					4	4	4	
Austroagrion watsoni	8	7		1						5	3	4	1
Austrocnemis splendida										1			
Ausrocoenagrion lyelli												1	
Ischnura aurora	1	2	1							2	1		
Ischnura heterosticta	3	1	2	1			1		1	2	55	1	3
Pseudagrion aureofrons	3	4	2								28	1	1
Pseudagrion ignifer	1	9	1	2						1	17	6	5
Pseudagrion microcephalum		2		1						1	1	1	
Xanthagrion erythroneurum		1	1										2
Family Aeshnidae													
Acanthaeschna victoria												3	
Adversaeschna brevistyla	1	3								2	1	1	2
Antipodophlebia asthenes	1												1
Austroaeschna parvistigma	5	4		1							2	5	2
Austroaeschna pulchra	4	4	2	4	1			1				4	6
Austroaeschna sigma	16	5	5							1	3	7	21
Austroaeschna subapicalis	6	14									2	4	5
Austroaeschna unicornis		2	1	1					1				1
Austrophlebia costalis	1									1	4	4	3
Dendroaeschna conspersa					1								1
Hemianax papuensis	2	4	2	3						1	3	11	
Notoaeschna geminata	3			1					1	5	6	3	
Spinaeschna tripunctata	4									1	5	3	
Telephlebia cyclops	2											1	1

	J	F	M	A	M	J	J	A	S	O	N	D	n.d.
Telephlebia godeffroyi	7										1	2	1
Family Petaluridae													
Petalura gigantea	6												2
Petalura litorea	1												
Family Gomphidae													
Austrogomphus amphiclitus	6	2	2	1						2	7	11	1
Austrogomphus cornutus	1		1	1							1	1	
Austrogomphus guerini	6									1	5	5	1
Austrogomphus ochraceus	23	6	2	1						1	15	20	2
Hemigomphus gouldii	4	3								1	4	8	
Hemigomphus heteroclytus	9	3	2							2	33	6	
Ictinogomphus australis	5												
Family Synthemistidae													
Archaeosynthemis orientalis	10		2								1	1	
Choristhemis flavoterminata	9	4			1						3	4	
Eusynthemis aurolineata	6	13								2		17	14
Eusynthemis brevistyla		2									1	11	4
Eusynthemis nigra											1		
Eusynthemis rentziana	1										1	6	2
Eusynthemis tillyardi											1		
Eusynthemis ursa											2	1	
Eusynthemis ursula											5	5	
Eusynthemis virgula	2	2								1	3	5	
Parasynthemis regina	1	1									1	1	
Synthemis eustalacta	10	7	1							1	6	11	5
Tonyosynthemis ofarrelli	3												
Family Corduliidae													
Hemicordulia australiae	4	9	2	4						3	6	3	
Hemicordulia continentalis		5	2							1		1	
Hemicordulia intermedia	2										1	4	
Hemicordulia superba											1		
Hemicordulia tau	2	3	1	1							2	4	
Procordulia jacksoniensis		1	1								1	3	2
Family Libellulidae													
Agrionoptera i.allogenes		1											
Austrothemis nigrescens	4	4								11		3	
Crocothemis nigrifrons	3	4	1								1	2	
Diplacodes bipunctata	14	16		9	3					4	8	5	1

	J	F	M	A	M	J	J	A	S	O	N	D	n.d.
Diplacodes haematodes	14	12	3	4	1					3	30	5	3
Diplacodes melanopsis	1	3		1	1					1	2	1	
Nannodiplax rubra	3	2	1	1	1					2	2		
Nannophlebia risi	11	5	2	3	1					9	4	6	
Nannophya australis	5	4	1							6		5	1
Nannophya dalei	11									2	7	7	2
Orthetrum boumiera	3	2		1									
Orthetrum caledonicum	13	5	1	1						1	4	4	1
Orthetrum sabina	3	9		2						2	2	1	
Orthetrum villosovittatum	10	12	1	2	3					3	2	4	1
Pantala flavescens		1		1									
Rhyothemis graphiptera	3	3									2	2	
Rhyothemis phyllis	1	5										1	
Tramea eurybia	1	2											
Tramea loewii	4	7	1	1					1	1	1	1	
Zyxomma elgneri	1												
Genera incertae sedis													
Austrocordulia refracta											2	1	
Cordulephya montana												1	
Cordulephya pygmaea			1	1									
Micromidia convergens				1									

MDD   Murray Darling Depression   19,958,349 ha

An extensive gently undulating sand and clay plain of Tertiary and Quaternary age frequently overlain by aeolian dunes. Vegetation consists of semi-arid woodlands of Black Oak / Belah, Bullock Bush/ Rosewood (*Alectryon oleifolius*) and *Acacia* spp., mallee shrublands and heathlands and savanna woodlands.

The region is known in Victoria as the Victorian Mallee region and characteristically has few surface water bodies

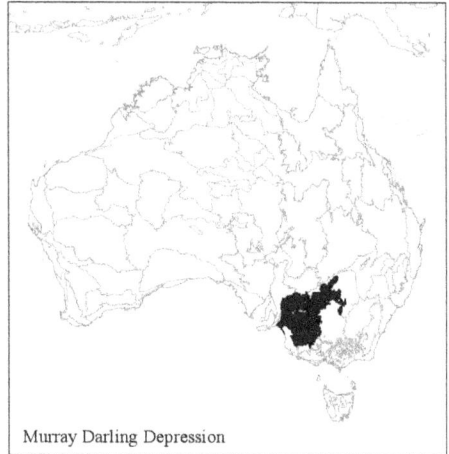

Murray Darling Depression

because its soils are highly permeable and its climate promotes high evaporative losses. Approximately 70 per cent of Victoria's mallee vegetation has been cleared and as a direct consequence of farming practices, the 1930s saw a part of the Victorian Mallee become one of the worst wind eroded areas in Australia. Substantial areas of mallee remain today in the western aeolian dunes, mainly in South Australia and but also western NSW. Clearing has also been widespread in the north eastern portion of the bioregion in NSW particularly on the undulating plains and relict river channels and lakes associated with the Murray and Darling Rivers.

Species   29   Specimens   586   Adults   449   Larvae   137

	J	F	M	A	M	J	J	A	S	O	N	D	n.d.
Family Lestidae													
Austrolestes analis			1									4	1
Austrolestes annulosus	2	1	1			1			1	11	2	3	1
Austrolestes aridus	2	1							1	7		6	
Austrolestes leda	1		3						9	9		3	1
Lestes concinnus													
Family Argiolestidae													
Austroargiolestes icteromelas	2												

	J	F	M	A	M	J	J	A	S	O	N	D	n.d.
Family Isostictidae													
Rhadinosticta simplex		2											
Family Platycnemidae													
Nososticta solida		2									1		
Family Coenagrionidae													
Austroagrion watsoni		1									1		
Ischnura aurora	6	1	13			1	1		1	7	4	9	
Ischnura heterosticta	4	7	3	2	1				1	5	1	2	1
Pseudagrion aureofrons		1							3				
Xanthagrion erythroneurum	6	7	5	1	3	1			1	13	5	3	1
Family Aeshnidae													
Adversaeschna brevistyla		1								1	3	1	
Austroaeschna subapicalis	1												
Austroaeschna unicornis													1
Hemianax papuensis	3	4	4					1	2	4	3	3	
Family Gomphidae													
Austroepigomphus praeruptus													1
Austrogomphus australis	2	3									1	13	
Austrogomphus bifurcatus													
Family Corduliidae													
Hemicordulia australiae											1		
Hemicordulia tau	2	1	2	1	1	1			2	15	16	7	
Family Libellulidae													
Austrothemis nigrescens	1												
Crocothemis nigrifrons	1												
Diplacodes bipunctata	5	1	9		1	1	1		7	32	10	7	
Diplacodes haematodes		1											
Orthetrum caledonicum	5	4	6							4	3	3	
Orthetrum migratum													
Pantala flavescens			5										
Tramea loewii			1										

COP   Cobar Peneplain   7,385,346 ha

Undulating plains and low hills
on Palaeozoic rocks; earths,
lithosols; *Eucalyptus populnea*
and *E. intertexta* woodlands
with mulga (*Acacia aneura*) in
the more arid areas. Semi-arid
climate.

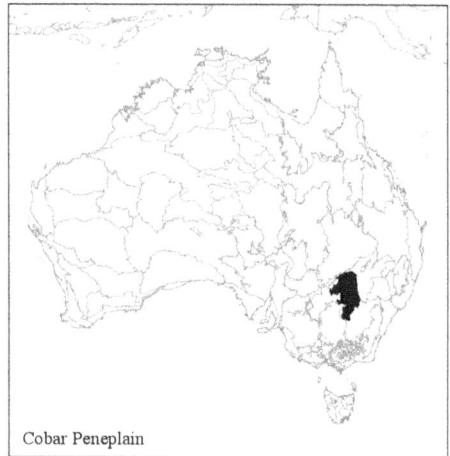

Cobar Peneplain

Species   19   Specimens   126   Adults   88   Larvae   38

	J	F	M	A	M	J	J	A	S	O	N	D	n.d.
Family Lestidae													
Austrolestes aridus			2										
Austrolestes leda			3										
Family Coenagrionidae													
Ischnura aurora	3		29	1									
Xanthagrion erythroneurum			4							3			
Family Aeshnidae													
Adversaeschna brevistyla			1										
Austrogynacantha heterogena			2										
Hemianax papuensis			3										
Family Gomphidae													
Austrogomphus australis	1		1								1	1	
Family Corduliidae													
Hemicordulia tau			1						1				
Family Libellulidae													
Diplacodes bipunctata			21	2									
Diplacodes haematodes			3										
Orthetrum caledonicum			5										

130

SYB   Sydney Basin   3,629,597 ha

Mesozoic sandstones and
shales; dissected plateaus;
forests, woodlands and heaths;
skeletal soils, sands and
podzolics.

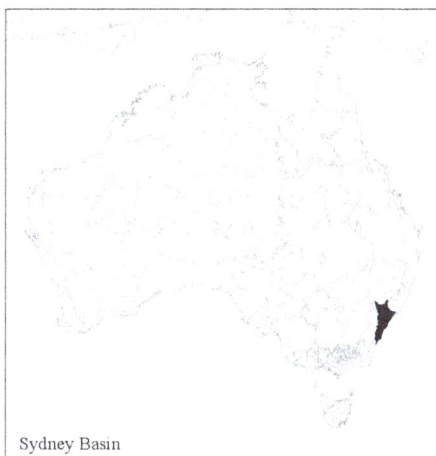

Sydney Basin

Species   86   Specimens   3147   Adults   3115   Larvae   32

	J	F	M	A	M	J	J	A	S	O	N	D	n.d.
Family Synlestidae													
Synlestes selysi		3	7	6	3					1	2	1	
Synlestes weyersii	19	7	7	10	6					1	9	21	8
Family Lestidae													
Austrolestes analis	12	6	2	2	2						5	1	
Austrolestes annulosus	2	3	1	1							3		1
Austrolestes aridus			1	3									
Austrolestes cingulatus	2	2	2	1						8	3	2	
Austrolestes io	1										1		
Austrolestes leda	18	4	7	3		2		6	14	12	11	19	3
Austrolestes psyche	14	8	3	7	2					2	15	7	
Family Argiolestidae													
Austroargiolestes icteromelas	25	13	11	6	2	1		1	8	7	33	50	8
Austroargiolestes isabellae	6							4	5	14	21	27	
Griseargiolestes griseus	38	17	4	2					1	4	21	32	23
Family Lestoideidae													
Diphlebia lestoides									5	11	10	17	1

	J	F	M	A	M	J	J	A	S	O	N	D	n.d.
Diphlebia nymphoides			1								5		
Family Isostictidae													
Neosticta canescens	2								1	1	6	15	17
Rhadinosticta simplex	9	5	11	3						2	7	11	1
Family Platycnemidae													
Nososticta solida	7	1	1							3	8	3	
Family Coenagrionidae													
Agriocnemis pygmaea	2	6								2	8	7	
Argiocnemis rubescens	3	2	1							3	1	1	1
Austroagrion watsoni	8	10	3	2					1	3	14	21	3
Austrocnemis splendida											1	1	
Caliagrion billinghursti				1							10	5	3
Ischnura aurora	16	3	2	1	1				1	3	4	13	4
Ischnura heterosticta	14	7	4	7	1			1	3	17	13	6	15
Pseudagrion aureofrons				1									
Pseudagrion microcephalum											6	2	1
Xanthagrion erythroneurum	5	5	2	2					1	2	5	5	2
Family Austropetaliidae													
Austropetalia patricia				1					1	16	11	2	4
Family Aeshnidae													
Adversaeschna brevistyla	16	10	5	1	1				2	7	9	16	5
Antipodophlebia asthenes												5	
Austroaeschna obscura	17	13	4	5	4					1	8	10	3
Austroaeschna parvistigma		2	1	4							2	2	
Austroaeschna pulchra	10	2	6	6							2	4	4
Austroaeschna sigma	1	1		2	1						1	1	
Austroaeschna subapicalis	8	9		2						1	1	5	1
Austroaeschna unicornis	4		2	9						2		1	4
Austrophlebia costalis	1	4	1						1	2	1	5	
Dendroaeschna conspersa			8	12	2				1			3	1
Hemianax papuensis	10	7	8	2			1	3	9	9	11	7	
Notoaeschna sagittata	4									3	5	5	2
Telephlebia godeffroyi	20	8	6	1						1	10	23	10
Family Petaluridae													
Petalura gigantea	186	27	1			1				2	43		
Family Gomphidae													
Austroepigomphus praeruptus	3		1							1	4	3	3
Austrogomphus amphiclitus	1	1	1									3	2

	J	F	M	A	M	J	J	A	S	O	N	D	n.d.
Austrogomphus cornutus		1	2	2							1	3	
Austrogomphus guerini	11	5		1						9	6	7	6
Austrogomphus ochraceus	11	10	4	2					1	1	16	39	9
Hemigomphus gouldii	7	5								1	4	33	10
Hemigomphus heteroclytus	3		1	1					1	1		1	2
Ictinogomphus australis													2
Family Synthemistidae													
Archaeosynthemis orientalis	20	3								1	6	7	1
Choristhemis flavoterminata	10	2	1	4	1						8	22	2
Eusynthemis aurolineata		1											
Eusynthemis brevistyla	1	2									4	1	1
Eusynthemis guttata												2	
Eusynthemis rentziana											1		
Eusynthemis tillyardi	16	7	1	1						4	16	27	5
Eusynthemis virgula	1				3				1		2	9	9
Parasynthemis regina	4		2								1	5	
Synthemis eustalacta	9	5	2	1							6	9	3
Family Corduliidae													
Hemicordulia australiae	16	17	11	10					3	4	13	19	1
Hemicordulia tau	7	5	3	3					2	11	10	10	
Procordulia jacksoniensis	4									1	2	5	
Family Libellulidae													
Crocothemis nigrifrons												1	1
Diplacodes bipunctata	11	11	6	15	2					6	5	11	1
Diplacodes haematodes	6	7	4	4						2	13	13	
Diplacodes melanopsis	3	2							1		2	3	1
Nannophlebia risi			1	1	1					1	1	5	
Nannophya australis	3	2									6	1	3
Nannophya dalei	7	2								1		22	
Orthetrum caledonicum	16	15	3	5						3	8	12	
Orthetrum sabina	2	4	3	1							1		
Orthetrum villosovittatum	12	5	4	4						1	4	6	
Pantala flavescens	1		4	1							1		3
Rhyothemis graphiptera	3										1	8	1
Tramea loewii	1	2	5						2	5	3		
Genera incertae sedis													
Archaeophya adamsi		1										3	2
Austrocordulia leonardi										1	8	4	1

	J	F	M	A	M	J	J	A	S	O	N	D	n.d.
Austrocordulia refracta									2	2	8	6	
Cordulephya divergens			1	1	1								
Cordulephya montana		2	1									1	2
Cordulephya pygmaea		1	16	15	5	3					1	1	3

RIV   Riverina   9,704,469 ha

An ancient riverine plain and alluvial fans composed of unconsolidated sediments with evidence of former stream channels. The Murray and Murrumbidgee Rivers and their major tributaries, the Lachlan and Goulburn Rivers flow westwards across this plain. Vegetation consists of river red gum and black box forests, box woodlands, saltbush shrublands, extensive grasslands and swamp communities.

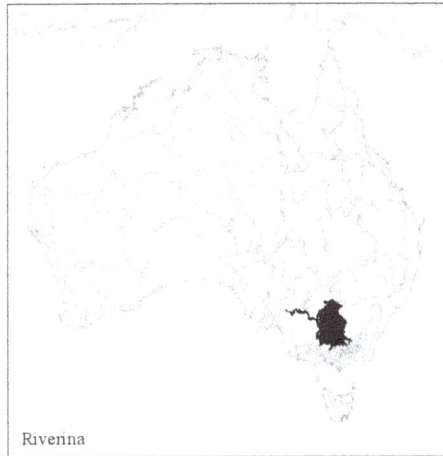

Riverina

Species   46   Specimens 1369   Adults   635   Larvae   734

	J	F	M	A	M	J	J	A	S	O	N	D	n.d.
Family Synlestidae													
Synlestes weyersii	1	1									1		
Family Lestidae													
Austrolestes analis	1	4	2	2							6	5	
Austrolestes annulosus	1	2	4	1						2	1	4	
Austrolestes aridus											1	1	
Austrolestes leda	7	3	3		1			1		1	3	8	
Family Argiolestidae													
Austroargiolestes calcaris											1		
Austroargiolestes icteromelas	2	1										3	
Family Lestoideidae													
Diphlebia nymphoides	2	1											
Family Isostictidae													
Rhadinosticta simplex	1		1										
Family Platycnemidae													
Nososticta solida	18	14	4									10	
Family Coenagrionidae													
Austroagrion watsoni		4	1							2	1	1	
Ischnura aurora	8	14	6	4				2	2	5	2	4	1

	J	F	M	A	M	J	J	A	S	O	N	D	n.d.
Ischnura heterosticta	16	19	5	4					1	2	5	11	
Pseudagrion aureofrons		9								1		2	
Xanthagrion erythroneurum	7	25	13	5						5	2	23	2
Family Aeshnidae													
Adversaeschna brevistyla		3	1									2	
Austroaeschna multipunctata		1									1	1	
Austroaeschna unicornis	2	1		1									
Hemianax papuensis	5	2	3							3	1	1	2
Notoaeschna sagittata		1											
Telephlebia brevicauda	1	1											
Family Gomphidae													
Antipodogomphus acolythus											1		
Austroepigomphus praeruptus	3	1											
Austrogomphus angelorum	1										2	2	
Austrogomphus australis	8	2	3							5		16	2
Austrogomphus cornutus	2		1									3	
Austrogomphus guerini	1	4											
Austrogomphus ochraceus	5		2								1	4	
Hemigomphus gouldii	4	1	1								1	1	
Family Synthemistidae													
Eusynthemis brevistyla		2											
Eusynthemis guttata	1												
Eusynthemis virgula	1									1	1	2	
Parasynthemis regina	6	6										1	
Family Corduliidae													
Hemicordulia australiae	2	2										1	
Hemicordulia tau	4	7	4	3	1				3	3	3	1	1
Family Libellulidae													
Diplacodes bipunctata	10	20	11	6						5	4	8	
Diplacodes haematodes		4								1			
Diplacodes melanopsis	2	1								1	1	4	
Nannophlebia risi	7												
Orthetrum caledonicum	7	21	8	1						2	1	3	
Genera incertae sedis													
Apocordulia macrops												1	
Cordulephya pygmaea			1										

NSS   NSW South Western Slopes   8,681,126 ha

An extensive area of foothills and isolated ranges comprising the lower inland slopes of the Great Dividing Range extending through southern New South Wales to western Victoria. Vegetation consists of wet/damp sclerophyll forests, peppermint forests and box/ironbark woodlands. Extensively cleared for agriculture.

NSW South Western Slopes

Species   56   Specimens   950   Adults   306   Larvae   644

	J	F	M	A	M	J	J	A	S	O	N	D	n.d.
Family Synlestidae													
Synlestes weyersii	4	3	1	1									
Family Lestidae													
Austrolestes analis	1									1	5	6	
Austrolestes annulosus	5							1				1	
Austrolestes aridus									1				
Austrolestes cingulatus	1												
Austrolestes leda	6	1		1				2		5		1	
Austrolestes psyche											2		
Family Argiolestidae													
Austroargiolestes icteromelas	10	5								1	2	2	
Griseargiolestes intermedius	1												
Family Lestoideidae													
Diphlebia lestoides												1	
Diphlebia nymphoides										1			
Family Platycnemidae													
Nososticta solida			1									1	
Family Coenagrionidae													
Austroagrion watsoni	2	1								1	2		

	J	F	M	A	M	J	J	A	S	O	N	D	n.d.
Ischnura aurora		3	2	1						5		4	
Ischnura heterosticta	3	5	3									2	
Pseudagrion aureofrons		1										1	
Xanthagrion erythroneurum	6	3										1	
Family Aeshnidae													
Adversaeschna brevistyla	1			2							1		
Austroaeschna inermis			1										
Austroaeschna multipunctata	1	2											
Austroaeschna pulchra	1	3	1										1
Austroaeschna unicornis	1	2											
Austrogynacantha heterogena		1											1
Hemianax papuensis			4	1					2	2		1	1
Notoaeschna sagittata		3											
Spinaeschna tripunctata										1			
Telephlebia brevicauda		1											
Family Gomphidae													
Austrogomphus amphiclitus		1											
Austrogomphus cornutus												1	1
Austrogomphus guerini	2	4	1								1		
Austrogomphus ochraceus	2	1											
Hemigomphus gouldii	3												
Hemigomphus heteroclytus			2									2	
Family Synthemistidae													
Eusynthemis brevistyla	3	1											
Eusynthemis guttata	2	1											
Eusynthemis virgula	4												
Parasynthemis regina	4	1											
Synthemis eustalacta	6												
Family Corduliidae													
Hemicordulia australiae	6										1		
Hemicordulia tau	3	1	3							5	3		
Family Libellulidae													
Diplacodes bipunctata	5	2	1	2					5	8	6	5	
Diplacodes haematodes	1	3	1								3	1	
Diplacodes melanopsis	3												
Nannophya dalei										1			
Orthetrum caledonicum	6	6	2							1	1	1	
Orthetrum villosovittatum	1												1

	J	F	M	A	M	J	J	A	S	O	N	D	n.d.
Genera incertae sedis													
Apocordulia macrops	15												2
Austrocordulia refracta	1												

SEH   South Eastern Highlands   8,375,961 ha

Steep dissected and rugged ranges extending across southern and eastern Victoria and southern NSW. Geology predominantly Palaeozoic rocks and Mesozoic rocks. Vegetation predominantly wet and dry sclerophyll forests, woodland, minor cool temperate rainforest and minor grassland and herbaceous communities. Large areas, particularly in the Box-Ironbark Forests, were felled for fuel and timber for the mines during the gold rushes in Victoria. Large areas have also been cleared in NSW for grazing or plantations.

South Eastern Highlands

Species   78   Specimens   5757   Adults   5072   Larvae   685

	J	F	M	A	M	J	J	A	S	O	N	D	n.d.
Family Hemiphlebiidae													
Hemiphlebia mirabilis	3	1									3		2
Family Synlestidae													
Synlestes weyersii	100	114	43	24	4				2	3	15	58	1
Family Lestidae													
Austrolestes analis	43	18	5	3							20	36	
Austrolestes annulosus	5	7	8	1					1	1	5	1	
Austrolestes aridus							1				6		
Austrolestes cingulatus	22	20	5							1	17	41	
Austrolestes io				1		1				6		1	
Austrolestes leda	15	8	6	1				9	6	25	26	37	
Austrolestes psyche	4	3	2	1					1	7	15	9	
Family Argiolestidae													
Austroargiolestes calcaris	31	6	2							9	31	80	
Austroargiolestes icteromelas	99	77	9	1					2	26	71	172	2
Griseargiolestes eboracus	11										3	8	1

	J	F	M	A	M	J	J	A	S	O	N	D	n.d.
Griseargiolestes griseus	26	1									8	3	
Griseargiolestes intermedius	10		2							1	3	15	1
Family Lestoideidae													
Diphlebia lestoides	22	1								1	30	84	
Diphlebia nymphoides	35	7									5	24	3
Family Isostictidae													
Neosticta canescens	2											2	
Rhadinosticta simplex	5	9	9	5								3	
Family Platycnemidae													
Nososticta solida	11	6										13	
Family Coenagrionidae													
Agriocnemis pygmaea												1	
Austroagrion watsoni	48	25	12	2						2	13	35	1
Austrocnemis splendida	1	3								2	6		3
Ausrocoenagrion lyelli	1										9	8	
Caliagrion billinghursti												1	
Ischnura aurora	13	11	8	2						16	24	24	
Ischnura heterosticta	34	30	10	4						10	24	44	2
Pseudagrion aureofrons			10	1								6	
Pseudagrion ignifer	1												
Xanthagrion erythroneurum	8	10	9	2						1	5	20	1
Family Austropetaliidae													
Austropetalia tonyana			1						1	6	18	3	2
Family Aeshnidae													
Adversaeschna brevistyla	31	12	10	2					2	9	30	58	1
Austroaeschn atrata	10	7	4	1						1	10		3
Austroaeschna flavomaculata	1									1			
Austroaeschna inermis	3	5	13	2						1			1
Austroaeschna multipunctata	14	60	29	13	5					2	7		23
Austroaeschna obscura	3		1							2	5		1
Austroaeschna parvistigma	15	5	4	1						2	9		
Austroaeschna pulchra	24	10	31	9						1	3	10	1
Austroaeschna subapicalis	19	10	8	3						1	3		2
Austroaeschna unicornis	22	16	28	9						1	5		2
Austrophlebia costalis											1		
Dendroaeschna conspersa			1								1	2	
Hemianax papuensis	17	8	12	2				4	4	4	2	12	
Notoaeschna sagittata	18	6	1								7	32	1

	J	F	M	A	M	J	J	A	S	O	N	D	n.d.
Spinaeschna tripunctata	10	1									1	4	
Telephlebia brevicauda	45	17	11	3							1	18	3
Telephlebia godeffroyi	6		2	1								4	
Family Petaluridae													
Petalura gigantea	4	3										14	2
Family Gomphidae													
Austrogomphus amphiclitus											1		
Austrogomphus cornutus	4	3									1	4	1
Austrogomphus guerini	101	41	15	1							24	70	2
Austrogomphus ochraceus	19	6	2	1							4	5	
Hemigomphus gouldii	46	9	3	1					1		7	30	2
Hemigomphus heteroclytus	3		1									10	
Family Synthemistidae													
Archaeosynthemis orientalis	6	2	4									3	1
Eusynthemis brevistyla	82	7									16	89	1
Eusynthemis guttata	50	42	40	8							11	60	
Eusynthemis tillyardi	5	2										11	1
Eusynthemis virgula	27	6	4								6	16	2
Parasynthemis regina	2	2	2									1	
Synthemis eustalacta	44	29	13	1							3	43	1
Family Corduliidae													
Hemicordulia australiae	18	14	7	2	1						7	26	
Hemicordulia superba												2	
Hemicordulia tau	38	18	21	5	1				2	18	24	70	
Procordulia jacksoniensis	10	2								1	1	6	
Family Libellulidae													
Diplacodes bipunctata	30	20	13	6					13	28	26	19	
Diplacodes haematodes	3	2	3							2	1	3	1
Diplacodes melanopsis	13	11	6	2							7	18	
Nannophlebia risi												1	
Nannophya dalei	21	11	1							3	9	11	
Orthetrum caledonicum	54	31	20	1							13	42	
Orthetrum villosovittatum	1	2											
Pantala flavescens	1												
Tramea loewii	1	1	1								1		
Genera incertae sedis													
Cordulephya montana	2	2										1	
Cordulephya pygmaea	2	5	14	6	3								1

SEC   South East Corner   2,532,053 ha

A series of deeply dissected near coastal ranges composed of Devonian granites and Palaeozoic sediments, inland of a series of gently undulating terraces (piedmont downs) composed of Tertiary sediments and flanked by Quaternary coastal plains, dune fields and inlets. The regional climate is strongly influenced by the Tasman Sea and the close proximity of the coast to the Great Dividing Range. Vegetation consists of high elevation woodlands,

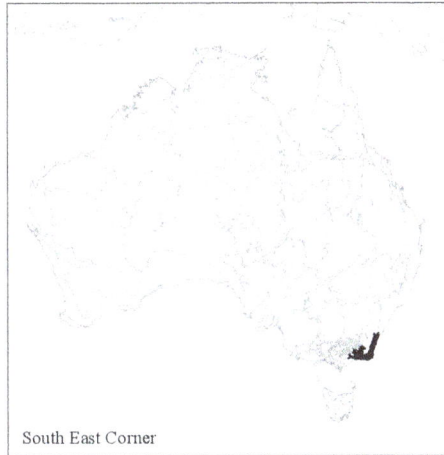

South East Corner

wet and damp sclerophyll forests interspersed with rain-shadow woodlands in the Snowy River Valley. Lowland and coastal sclerophyll forests, woodlands, warm temperate rainforest and coastal communities occur in the lower areas.

Species   71   Specimens   1230   Adults   1119   Larvae   111

	J	F	M	A	M	J	J	A	S	O	N	D	n.d.
Family Synlestidae													
Synlestes selysi		1		5									
Synlestes weyersii	44	15	10	5		1					4	14	
Family Lestidae													
Austrolestes analis	7	8								1	7	6	
Austrolestes annulosus				1									
Austrolestes cingulatus	12	4	2	1						2	4	4	
Austrolestes io		1								1		2	
Austrolestes leda	5									2	9	3	1
Austrolestes psyche	5	4								2	3	3	
Family Argiolestidae													
Austroargiolestes calcaris	1										5		
Austroargiolestes icteromelas	45	20	6	4							10	14	
Griseargiolestes griseus											1		3

	J	F	M	A	M	J	J	A	S	O	N	D	n.d.
Family Lestoideidae													
Diphlebia lestoides	1										15	6	
Diphlebia nymphoides	7	6									1	2	
Family Isostictidae													
Rhadinosticta simplex	4	5	6	2									
Family Platycnemidae													
Nososticta solida	4	1										2	
Family Coenagrionidae													
Argiocnemis rubescens	1												
Austroagrion watsoni	23	9	3	4						3	6	9	2
Austrocnemis splendida	3	1								1	1		
Ausrocoenagrion lyelli		1									5		
Caliagrion billinghursti		1											
Ischnura aurora	6	5	2	1						2	11	10	
Ischnura heterosticta	13	6	3	3						2	4	5	
Pseudagrion aureofrons				1									
Pseudagrion microcephalum	4		1	1									
Xanthagrion erythroneurum	1	2											
Family Austropetaliidae													
Austropetalia tonyana										1			
Family Aeshnidae													
Acanthaeschna victoria													4
Adversaeschna brevistyla	16	6		3						3	8	1	
Austroaeschna inermis			2							1			
Austroaeschna multipunctata		1	4	1								1	
Austroaeschna obscura	1												
Austroaeschna pulchra	4	3	9	6						1			
Austroaeschna subapicalis	1		1										
Austroaeschna unicornis	7	11	5	2						1			1
Austrophlebia costalis	5	1	2								1		
Dendroaeschna conspersa			6	5									2
Hemianax papuensis	3	4	1					1	2	3			
Notoaeschna sagittata	1	3								1	4	3	
Spinaeschna tripunctata										3		3	
Telephlebia brevicauda	4												
Telephlebia godeffroyi	1	1											
Family Petaluridae													
Petalura gigantea												1	

	J	F	M	A	M	J	J	A	S	O	N	D	n.d.
Family Gomphidae													
Austrogomphus amphiclitus	3									1			
Austrogomphus cornutus										1			
Austrogomphus guerini	22	13	3	2							3	8	
Austrogomphus ochraceus	25	8	3	4							1	6	
Hemigomphus gouldii	22	12	2							1	1	2	
Hemigomphus heteroclytus	1												
Family Synthemistidae													
Archaeosynthemis orientalis	1												
Choristhemis flavoterminata	2		1										
Eusynthemis brevistyla	2	2										3	4
Eusynthemis guttata	2		1										
Eusynthemis tillyardi		1		1									
Eusynthemis virgula	3	7									4		1
Synthemis eustalacta	2	1								1			
Family Corduliidae													
Hemicordulia australiae	23	11	5	3							1	5	
Hemicordulia tau	1	1								1	5		
Procordulia jacksoniensis										1		1	
Family Libellulidae													
Diplacodes bipunctata	10	7	4	2						3	4	4	
Diplacodes haematodes	4	4	5	6							2	1	
Diplacodes melanopsis	9	5										2	
Nannophlebia risi			1										1
Nannophya australis	6		1										
Nannophya dalei	2										2	1	
Orthetrum caledonicum	13	8	5	3							3		1
Orthetrum villosovittatum	13	6	1										1
Tramea loewii	4												
Genera incertae sedis													
Austrocordulia refracta	3									2	4	3	
Cordulephya montana	2		1								1	3	1
Cordulephya pygmaea	2	1	7	1		2							1

AUA Australian Alps 1,232,981 ha

A series of high elevation plateaux capping the South Eastern Highlands (Region SEH) and the southern tablelands in NSW. The geology consists largely of granitic and basaltic rocks. Vegetation is dominated by alpine herbfields, and other treeless communities, snow gum woodlands and montane forests dominated by alpine ash. The Victorian Alps region essentially is bounded by the 1200 metre contour.

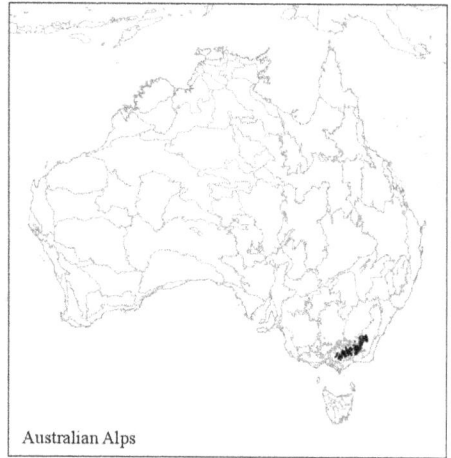

Australian Alps

Species 46 Specimens 827 Adults 773 Larvae 54

	J	F	M	A	M	J	J	A	S	O	N	D	n.d.
Family Synlestidae													
Synlestes weyersii	5	9	4										
Family Lestidae													
Austrolestes analis	4	5	1										
Austrolestes annulosus			1										
Austrolestes cingulatus	6	18	3									4	
Austrolestes leda	4	6	2							2	1	2	
Austrolestes psyche	4	6	3									2	
Family Argiolestidae													
Austroargiolestes calcaris	34	11	1							1	1	23	1
Austroargiolestes icteromelas	5	5	1									7	
Griseargiolestes intermedius	8	34									1	4	
Family Lestoideidae													
Diphlebia lestoides												2	
Family Coenagrionidae													
Austroagrion watsoni	5	3										1	
Ausrocoenagrion lyelli		4											
Ischnura aurora	7	1	5								1	2	
Ischnura heterosticta	2	4										1	

	J	F	M	A	M	J	J	A	S	O	N	D	n.d.
Xanthagrion erythroneurum	3	2	1										
Family Austropetaliidae													
Austropetalia tonyana											8		1
Family Aeshnidae													
Acanthaeschna victoria											2		
Adversaeschna brevistyla	4	2	1	1							1	2	
Austroaeschn atrata	11	3	3									1	
Austroaeschna flavomaculata	12	25	5										
Austroaeschna inermis	7	9	2	1							2		1
Austroaeschna multipunctata	4	10	5										
Austroaeschna parvistigma		2	1										
Austroaeschna pulchra	1	2											
Austroaeschna subapicalis		3	1										
Austroaeschna unicornis		1	1										1
Hemianax papuensis	5	2	3								2	1	
Notoaeschna sagittata	2		1										
Telephlebia brevicauda	15	18	4			1			1		2	1	1
Family Gomphidae													
Austrogomphus guerini	5	4	1							1		6	
Austrogomphus ochraceus			1										
Hemigomphus gouldii	2											5	
Hemigomphus heteroclytus												1	
Family Synthemistidae													
Archaeosynthemis orientalis	1	1											
Eusynthemis brevistyla	14	5										7	
Eusynthemis guttata	21	18	1									6	
Eusynthemis tillyardi												1	
Eusynthemis virgula	1											1	
Synthemis eustalacta	34	52	8								2	9	
Family Corduliidae													
Hemicordulia tau	7	3	1								1	4	
Procordulia jacksoniensis	4	6	1								1	4	
Family Libellulidae													
Diplacodes bipunctata	9	6									1	1	
Nannophya australis	1	1											
Nannophya dalei	6	32	1								1	3	
Orthetrum caledonicum			1										

VIM Victorian Midlands 3,469,789 ha

An extensive area of foothills and isolated ranges comprising the lower inland slopes of the Great Dividing Range extending from North-eastern Victoria to Casterton in Western Victoria. Large areas of the region were cleared during the gold rushes of the late nineteenth and early twentieth centuries so today it is characterised by patches of woodland and forest interspersed with a rural landscape with modified pastures and some cropping.

Victorian Midlands

Vegetation includes most of the Box Ironbark Woodland in Victoria, as well as substantial areas of Eucalyptus forests and woodlands with a grassy ground layer.

The flatter and more fertile areas of the Victorian Midlands have been substantially cleared for agriculture, principally sheep and beef cattle grazing. Timber harvesting remains an important land use in the Victorian Midlands. Much of the forests were extensively cut for timber to meet the demands of the gold mining industry of last century. In the less fertile parts of the Victorian Midlands, substantial areas of native vegetation remain today in good condition, for example, the Grampians National Park.

Species 61 Specimens 1849 Adults 1361 Larvae 488

	J	F	M	A	M	J	J	A	S	O	N	D	n.d.
Family Hemiphlebiidae													
Hemiphlebia mirabilis	6										5	48	38
Family Synlestidae													
Synlestes weyersii	6	2	5	2						1		4	
Family Lestidae													
Austrolestes analis	9	2	9							1	28	51	
Austrolestes annulosus	1	2	2						2	4	9	7	
Austrolestes aridus									2	4		8	

	J	F	M	A	M	J	J	A	S	O	N	D	n.d.
Austrolestes cingulatus	1											2	
Austrolestes io									1	3	3		
Austrolestes leda	5		2					2	17	15	21	44	2
Austrolestes psyche	2										3	3	
Family Argiolestidae													
Austroargiolestes calcaris	2										1	1	
Austroargiolestes icteromelas	6	3	3						1	2	19	22	2
Family Lestoideidae													
Diphlebia lestoides											1		1
Diphlebia nymphoides										1		1	1
Family Platycnemidae													
Nososticta solida	1											3	1
Family Coenagrionidae													
Austroagrion watsoni	3		2								8	21	2
Austrocnemis splendida	1										1		
Ausrocoenagrion lyelli											8		
Caliagrion billinghursti											5	5	
Ischnura aurora	7	4	4						6	17	33	42	
Ischnura heterosticta	9	6	9						2	4	15	25	
Pseudagrion aureofrons	1												
Xanthagrion erythroneurum	1	3	2						1	8	13	13	
Family Austropetaliidae													
Austropetalia tonyana											2		
Family Aeshnidae													
Adversaeschna brevistyla	7		2							1	7	20	1
Austroaeschn atrata													1
Austroaeschna ingrid	20		3									5	24
Austroaeschna multipunctata	2	1	2								2		
Austroaeschna parvistigma	2											1	
Austroaeschna pulchra	1		3	2									
Austroaeschna subapicalis	27	2		1		1						1	
Austroaeschna unicornis		2	4	2									
Hemianax papuensis	2	5	2					2	4	9	2	10	
Notoaeschna sagittata												1	
Spinaeschna tripunctata											1		
Telephlebia brevicauda	6					1					1	2	
Family Gomphidae													
Austrogomphus guerini	5	2	2								4	16	1

	J	F	M	A	M	J	J	A	S	O	N	D	n.d.
Austrogomphus ochraceus	1		1								2	8	2
Hemigomphus gouldii	2											1	
Hemigomphus heteroclytus			2										
Family Synthemistidae													
Archaeosynthemis orientalis												1	
Eusynthemis brevistyla	5	13									4	4	
Eusynthemis guttata	2	2									4	8	1
Eusynthemis tillyardi												1	
Eusynthemis virgula		2									3	6	
Parasynthemis regina											1		
Synthemis eustalacta	1	2	1								1	2	
Family Corduliidae													
Hemicordulia australiae												2	
Hemicordulia tau	5	4	7	3		1			10	14	17	27	2
Procordulia jacksoniensis												1	
Family Libellulidae													
Diplacodes bipunctata	10	5	6					2	31	31	25	32	1
Diplacodes haematodes	2	2	1	1							2	1	
Diplacodes melanopsis	1		8									8	4
Nannophya dalei										2		2	
Orthetrum caledonicum	6	4	3	1						4	3	18	
Tramea loewii											1	1	
Genera incertae sedis													
Cordulephya pygmaea		1		1								2	

SCP   South East Coastal Plain 1,749,237 ha

Undulating Tertiary and Quaternary coastal plains and hinterlands occur in several distinct segments (Warrnambool Plain, Otway Plain and Gippsland Plain) rise up to 200 metres in altitude and extend from Tyrendarra in the west to Lakes Entrance in the east and including Geelong, eastern Melbourne and the Mornington Peninsula. The area has a temperate climate with rainfall varying from about 500 to 1100 mm, typically with higher rainfall in winter. Adjacent areas of higher altitude (e.g. the Otway and Strzelecki Ranges) produce rainshadow effects in some parts of the area.

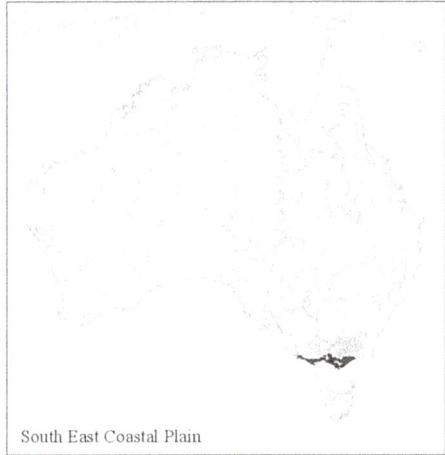
South East Coastal Plain

The Warrnambool Plain is dominated by nutrient deficient soils over low calcareous dune formations and the distinctive cliffed coastline. Much of the limestone has been overlain by more recent sediments, and between the limestone dunes, areas of swamplands are characterised by highly fertile peats and seasonal inundation. The area east of Warrnambool is characterised by deeper soils of volcanic origins overlying limestone, which are dissected by streams. The Otway Plain includes coastal plains, river valleys and foothills from the Bellarine Peninsula west to Princetown. A small isolated component at Werribee, on the western shore of Port Phillip Bay, is included. The Gippsland Plain includes lowland coastal and alluvial plains characterised by generally flat to gently undulating terrain. The coastline is varied and includes sandy beaches backed by dunes and cliffs, and shallow inlets with extensive mud and sand flats.

The vegetation includes lowland forests, open forests with shrubby or heathy understoreys, grasslands and grassy woodlands, heathlands, shrublands, freshwater and coastal wetlands, mangrove scrubs, saltmarshes, dune scrubs and coastal tussock grasslands. Extensively cleared for agriculture.

Species   48   Specimens   1055   Adults   1051   Larvae   4

151

	J	F	M	A	M	J	J	A	S	O	N	D	n.d.
Family Hemiphlebiidae													
Hemiphlebia mirabilis	10											1	
Family Synlestidae													
Synlestes weyersii			2	1							1	3	
Family Lestidae													
Austrolestes analis	23	6	3	2	1					1	15	32	
Austrolestes annulosus	17	6	1	1	1					2	1	21	
Austrolestes cingulatus		1	1	1						1			
Austrolestes io											1	4	
Austrolestes leda	8	3	4	3	2				7	6	12	21	2
Austrolestes psyche	5	1							2		3	6	
Family Argiolestidae													
Austroargiolestes icteromelas	5	2	2						4	3	6	17	
Griseargiolestes intermedius													3
Family Lestoideidae													
Diphlebia nymphoides													1
Family Isostictidae													
Rhadinosticta simplex	2	1		1									
Family Platycnemidae													
Nososticta solida	4		1									1	
Family Coenagrionidae													
Austroagrion watsoni	12	3	5	1						1	6	14	
Austrocnemis splendida													1
Ausrocoenagrion lyelli										1			
Ischnura aurora	17		3	1	1					6	18	13	
Ischnura heterosticta	28	21	8	1						2	12	25	1
Xanthagrion erythroneurum	28	3	7		1					2	10	16	
Family Austropetaliidae													
Austropetalia tonyana											2		
Family Aeshnidae													
Adversaeschna brevistyla	15	4	4							2	11	20	
Austroaeschna parvistigma		1											
Austroaeschna pulchra												1	
Austroaeschna unicornis	1	2	3	1								1	
Austrophlebia costalis													1
Hemianax papuensis	7	4	3	1					3	2	2	1	
Spinaeschna tripunctata												1	
Telephlebia brevicauda	1	1				1							

	J	F	M	A	M	J	J	A	S	O	N	D	n.d.
Family Gomphidae													
Austrogomphus guerini	1	2	1								1	1	
Hemigomphus gouldii		1											
Family Synthemistidae													
Eusynthemis brevistyla	2												
Eusynthemis guttata		1											
Parasynthemis regina												2	
Synthemis eustalacta	3	2									1	6	
Family Corduliidae													
Hemicordulia australiae	12	5	3								3	14	
Hemicordulia tau	24	9	5	4	2				3	8	17	24	1
Procordulia jacksoniensis	1	1									2	1	
Family Libellulidae													
Austrothemis nigrescens	1												
Crocothemis nigrifrons											1		
Diplacodes bipunctata	21	7	11	1	2				11	15	15	17	
Diplacodes haematodes	1	2		1								2	
Diplacodes melanopsis	13	24	1								1		1
Nannophya dalei	3										9	2	
Orthetrum caledonicum	24	6	6	2							3	15	
Pantala flavescens					1								
Tramea loewii			1										
Genera incertae sedis													
Cordulephya pygmaea	1												4

SVP   Southern Volcanic Plain   2,440,340 ha

An extensive undulating basaltic plain in south-western Victoria, stretching from Melbourne west to Portland, south to Colac and north to Beaufort. It is characterised by vast open areas of grasslands, small patches of open woodland, stony rises denoting old lava flows, the low peaks of long extinct volcanoes dotting the landscape and numerous scattered large shallow lakes with extensive wetlands.

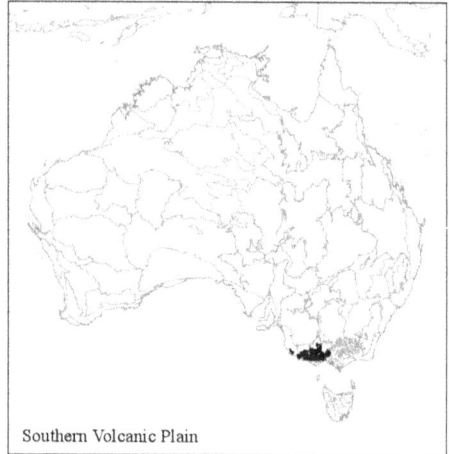

Southern Volcanic Plain

The grassland communities are floristically rich, usually dominated by Kangaroo Grass with a wide variety of perennial herbs. The open and fertile grassy plains were one of the first areas settled for agriculture in Victoria and native grasslands are now reduced to a few thousand hectares in extent. The major land use is agriculture, especially sheep and cattle grazing and cropping.

Species   37   Specimens   476   Adults   389   Larvae   87

	J	F	M	A	M	J	J	A	S	O	N	D	n.d.
Family Hemiphlebiidae													
Hemiphlebia mirabilis												2	1
Family Synlestidae													
Synlestes weyersii	1	1										2	
Family Lestidae													
Austrolestes analis	4		3							1	3	14	5
Austrolestes annulosus	1					1			4	17	3	10	
Austrolestes cingulatus	2		1								1	10	
Austrolestes io			1						1				
Austrolestes leda	2		1						4	3	2	4	2
Austrolestes psyche			2								1	3	
Family Argiolestidae													
Austroargiolestes icteromelas	2											7	3

	J	F	M	A	M	J	J	A	S	O	N	D	n.d.
Family Lestoideidae													
Diphlebia lestoides													1
Family Platycnemidae													
Nososticta solida		1											
Family Coenagrionidae													
Austroagrion watsoni			2									4	
Ausrocoenagrion lyelli	1	1										6	
Ischnura aurora	2		4							8	2	15	
Ischnura heterosticta	8	2	1							2	2	12	
Xanthagrion erythroneurum			1							25	2	10	
Family Aeshnidae													
Adversaeschna brevistyla	1	1	2							1	3	3	1
Austroaeschna multipunctata			4										
Austroaeschna pulchra	1												
Dendroaeschna conspersa				1									
Hemianax papuensis		1						1			3	5	1
Telephlebia brevicauda	1												
Family Gomphidae													
Austrogomphus guerini	9	1									2	13	
Hemigomphus gouldii												1	
Family Synthemistidae													
Synthemis eustalacta		2										2	
Family Corduliidae													
Hemicordulia australiae			4									9	
Hemicordulia tau	1		3							2	2	5	
Procordulia jacksoniensis			1							1	1	1	
Family Libellulidae													
Crocothemis nigrifrons			1										
Diplacodes bipunctata		2	5						4	4	5	3	6
Diplacodes haematodes											1		
Diplacodes melanopsis												3	
Nannophya dalei												2	
Orthetrum caledonicum	3		1									8	1
Genera incertae sedis													
Cordulephya pygmaea		1											

NCP   Naracoorte Coastal Plain   2,458,215 ha

A broad coastal plain of Tertiary and Quaternary sediments with a regular series of calcareous sand ridges separated by inter-dune swales closed limestone depressions and young volcanoes at Mount Gambier. Vegetation is dominated by heathy woodlands and mallee shrublands with wet heaths in the inter-dune swales. Extensively cleared for agriculture.

Naracoorte Coastal Plain

Species   31   Specimens   511   Adults   511   Larvae   0

	J	F	M	A	M	J	J	A	S	O	N	D	n.d.
Family Hemiphlebiidae													
Hemiphlebia mirabilis	6		1								3	16	17
Family Lestidae													
Austrolestes analis	4	2	3	2							11	31	
Austrolestes annulosus	2		1							4	3	15	
Austrolestes io										3		1	
Austrolestes leda												10	
Austrolestes psyche	1		1							6	6	12	
Family Argiolestidae													
Austroargiolestes icteromelas												1	
Family Coenagrionidae													
Austroagrion cyane	1										1	8	
Austroagrion watsoni											1	10	
Ausrocoenagrion lyelli											1	1	
Ischnura aurora		3	1					1	1	7	11	26	
Ischnura heterosticta	5	2	4							3	10	23	
Xanthagrion erythroneurum			1						1	1	3	9	
Family Aeshnidae													
Adversaeschna brevistyla	3	2	1							1	12	19	

	J	F	M	A	M	J	J	A	S	O	N	D	n.d.
Austroaeschna parvistigma			1										
Hemianax papuensis		1	5	1				1	1	4		8	
Family Gomphidae													
Austrogomphus cornutus				1									
Austrogomphus guerini		1									1		
Family Synthemistidae													
Eusynthemis brevistyla											2		
Synthemis eustalacta	1		1								3	9	
Family Corduliidae													
Hemicordulia australiae												1	
Hemicordulia tau	1		2							5	10	7	
Procordulia jacksoniensis										3	4	8	1
Family Libellulidae													
Austrothemis nigrescens											2	8	8
Crocothemis nigrifrons	2		2									4	
Diplacodes bipunctata	1	1	2						4	10	6	23	
Diplacodes haematodes		1											
Diplacodes melanopsis											1	3	
Nannophya dalei											2	2	
Orthetrum caledonicum	2	1									2	4	
Pantala flavescens		1		1									

KAN   Kanmantoo   812,415 ha

Temperate, well defined
uplands of Cambrian and
Late Proterozoic marine
sediments, and a lateritized
surface becoming increasingly
dissected northwards, with
eucalypt open forests and
woodlands and heaths on
mottled yellow and ironstone
gravelly duplex soils in the
wetter areas, and *Eucalyptus
odorata* and drooping sheoak
on shallow rocky soils in drier
areas. Extensively cleared for
agriculture.

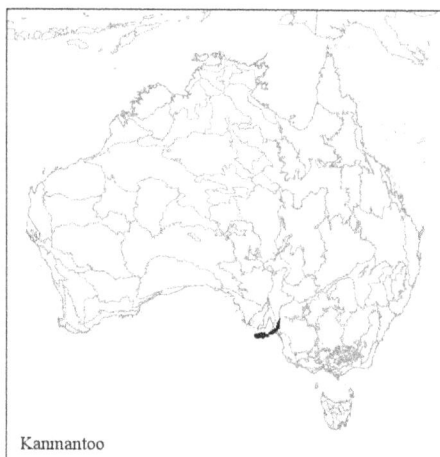

Kanmantoo

Species   17   Specimens   90   Adults   84   Larvae   6

	J	F	M	A	M	J	J	A	S	O	N	D	n.d.
Family Lestidae													
Austrolestes analis	6										4		
Austrolestes annulosus	9									3	8		
Austrolestes aridus											1		
Family Coenagrionidae													
Ischnura heterosticta											1		
Xanthagrion erythroneurum	4											1	
Family Aeshnidae													
Adversaeschna brevistyla											5	1	1
Agyrtacantha dirupta													
Austroaeschna parvistigma													2
Austroaeschna unicornis												1	
Hemianax papuensis	3	1											1
Family Gomphidae													
Austrogomphus guerini												1	
Family Synthemistidae													
Synthemis eustalacta												2	
Family Corduliidae													
Hemicordulia tau		1							1	2	8	1	1
Procordulia jacksoniensis	1										9		1

	J	F	M	A	M	J	J	A	S	O	N	D	n.d.
Family Libellulidae													
Austrothemis nigrescens											2		
Diplacodes haematodes										1			
Nannophya dalei	1												

FLB Flinders Lofty Block 6,615,765 ha

Temperate to arid Proterozoic ranges, alluvial fans and plains, and some outcropping volcanics, with the semi arid to arid north supporting native cypress, black oak (belah) and mallee open woodlands, Eremophila and Acacia shrublands, and bluebush/ saltbush chenopod shrublands on shallow, well-drained loams and moderately-deep, well-drained red duplex soils. The increase in rainfall to the south corresponds with an increase in low open woodlands of

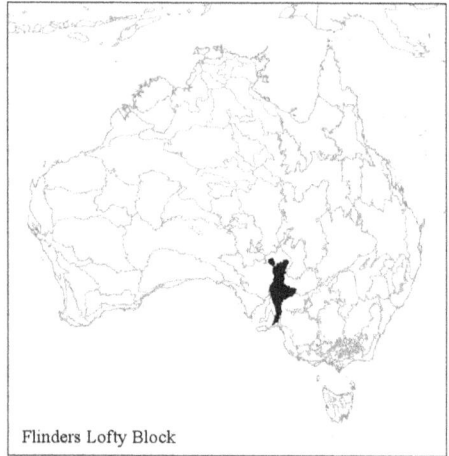

Flinders Lofty Block

Eucalyptus obliqua and *E. baxteri* on deep lateritic soils, and *E. fasciculosa* and *E. cosmophylla* on shallower or sandy soils.

Species   23   Specimens   193   Adults   192   Larvae   1

	J	F	M	A	M	J	J	A	S	O	N	D	n.d.
Family Lestidae													
Austrolestes analis		1								1	2		1
Austrolestes annulosus	2	1			2				1		2	2	
Austrolestes aridus					1					1			
Austrolestes psyche			1							1			2
Family Coenagrionidae													
Austroagrion cyane											1		
Austroagrion watsoni										2			
Ischnura aurora									1	3	2	2	
Ischnura heterosticta		1		1						1		1	1
Xanthagrion erythroneurum		3	1								4	2	
Family Aeshnidae													
Adversaeschna brevistyla	5	2		1					1	2	2	2	2
Austroaeschna parvistigma	1	1		1							5		
Austroaeschna unicornis	1	1	1	1			1						
Hemianax papuensis	1	1	3		1	1		1	1	1	1	4	4

	J	F	M	A	M	J	J	A	S	O	N	D	n.d.
Family Gomphidae													
Austroepigomphus praeruptus													1
Austrogomphus australis	1												
Austrogomphus guerini							1				1	7	1
Family Synthemistidae													
Synthemis eustalacta	5											4	1
Family Corduliidae													
Hemicordulia australiae	1												
Hemicordulia tau	3		6	8	1				4	6	1	5	3
Family Libellulidae													
Diplacodes bipunctata		1		1	3		1	1	1			2	1
Diplacodes haematodes		1		1	3	1				1	1	3	1
Orthetrum caledonicum	1			1	1				1	4		1	3
Pantala flavescens				1									

GAW   Gawler   12,002,883 ha

Semi arid to arid, flat topped
to broadly rounded hills of the
Gawler Range Volcanics and
Proterozoic sediments, low
plateaux on sandstone and
quartzite with an undulating
surface of aeolian sand or
gibbers and rocky quartzite
hills with colluvial footslopes,
erosional and depositional
plains and salt encrusted lake
beds, with black oak (belah)
and myall low open woodlands,
open mallee scrub, bluebush/
saltbush open chenopod
shrublands and tall mulga
shrublands on shallow loams, calcareous earths and hard red
duplex soils.

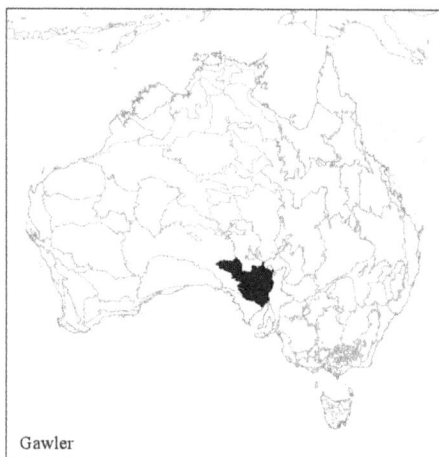

Gawler

Species   13   Specimens   27   Adults   27   Larvae   0

	J	F	M	A	M	J	J	A	S	O	N	D	n.d.
Family Lestidae													
Austrolestes annulosus				1									
Austrolestes aridus									1				
Family Coenagrionidae													
Ischnura aurora										1			
Xanthagrion erythroneurum										1	1		
Family Aeshnidae													
Adversaeschna brevistyla		1										1	
Austroaeschna parvistigma		1		3							1	3	
Austroaeschna unicornis		1											
Hemianax papuensis												1	
Family Synthemistidae													
Synthemis eustalacta		1											
Family Corduliidae													
Hemicordulia tau					1					1	1		1

	J	F	M	A	M	J	J	A	S	O	N	D	n.d.
Family Libellulidae													
Diplacodes bipunctata										1			
Diplacodes haematodes												1	
Orthetrum caledonicum										1	1	1	

SSD   Simpson Strzelecki Dunefields   27,984,283 ha

Arid dune fields and sandplains with sparse shrubland and spinifex hummock grassland, and cane grass on deep sands along dune crests. Large salt lakes, notably Lake Eyre and many clay pans are dispersed amongst the dunes. Several significant arid rivers terminate at Lake Eyre, Cooper Creek and Warburton River. They are fringed with coolibah and redgum woodlands.

Simpson Strzelecki Dunefields

Species   13   Specimens   94   Adults   94   Larvae   0

	J	F	M	A	M	J	J	A	S	O	N	D	n.d.
Family Lestidae													
Austrolestes annulosus									1		1		
Austrolestes aridus			3				2		1		3		
Austrolestes leda							1						
Family Coenagrionidae													
Ischnura aurora									2	2	1		
Ischnura heterosticta										2			
Xanthagrion erythroneurum		1	2						1	2	4		
Family Aeshnidae													
Austrogynacantha heterogena					1								
Hemianax papuensis			2	3	2		2		2		1		2
Family Corduliidae													
Hemicordulia tau	1		2	2	1		2		1		1		1
Family Libellulidae													
Diplacodes bipunctata			3		5			3		2	2		2
Diplacodes haematodes									1	1			
Orthetrum caledonicum			1	2	2			1		3	1	5	1
Pantala flavescens			1										

EYB   Eyre Yorke Block   6,120,409 ha

Archaean basement rocks and Proterozoic sandstones overlain by undulating to occasionally hilly calcarenite and calcrete plains and areas of aeolian quartz sands, with mallee woodlands, shrublands and heaths on calcareous earths, duplex soils and calcareous to shallow sands, now largely cleared for agriculture.

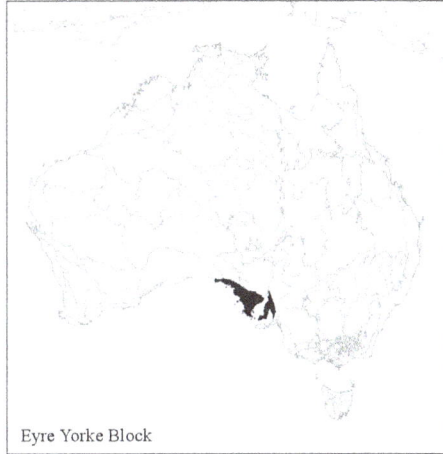

Eyre Yorke Block

Species   14   Specimens   43   Adults   43   Larvae   0

	J	F	M	A	M	J	J	A	S	O	N	D	n.d.
Family Lestidae													
Austrolestes analis					1								
Austrolestes annulosus				1							3	2	1
Austrolestes cingulatus								1					
Austrolestes leda													1
Family Coenagrionidae													
Ischnura aurora			1										
Ischnura heterosticta		2	1										1
Xanthagrion erythroneurum										1			
Family Aeshnidae													
Adversaeschna brevistyla										3			3
Hemianax papuensis		1	1								1		
Family Corduliidae													
Hemicordulia tau	1		1								3	2	
Procordulia jacksoniensis	1												
Family Libellulidae													
Diplacodes bipunctata								1	2	2			
Diplacodes haematodes									1				
Orthetrum caledonicum			1								1	1	1

ESP Esperance Plains 2,921,327 ha

Proteaceous Scrub and mallee heaths on sandplain overlying Eocene sediments; rich in endemics. Herbfields and heaths on abrupt granite and quartzite ranges that rise from the plain. Eucalypt woodlands occur in gullies and alluvial foot-slopes. Warm Mediterranean.

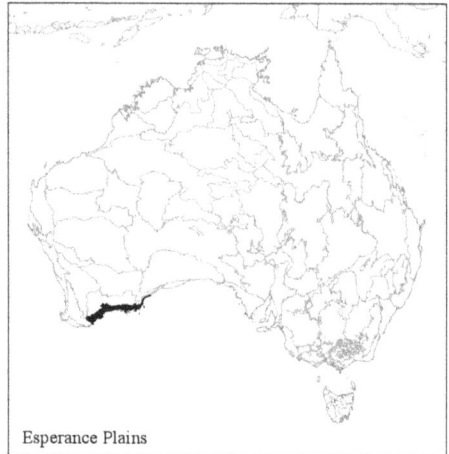

Esperance Plains

Species 20 Specimens 131 Adults 131 Larvae 0

	J	F	M	A	M	J	J	A	S	O	N	D	n.d.
Family Lestidae													
Austrolestes aleison		4		2								1	
Austrolestes analis	1	1		4							1	2	
Austrolestes annulosus				8						10	1	8	
Austrolestes io	1	1											
Family Argiolestidae													
Archiargiolestes pusillissimus												2	
Archiargiolestes pusillus												4	
Miniargiolestes minimus												3	
Family Coenagrionidae													
Austroagrion cyane				4						1	5	8	
Ischnura aurora				1									
Xanthagrion erythroneurum				2								2	
Family Aeshnidae													
Adversaeschna brevistyla		3		1								1	
Hemianax papuensis	2	1	1	2					1		1		
Family Gomphidae													
Austrogomphus collaris	1											4	

	J	F	M	A	M	J	J	A	S	O	N	D	n.d.
Family Synthemistidae													
Archaeosynthemis leachii												1	
Archaeosynthemis occidentalis	2												
Family Corduliidae													
Hemicordulia tau				8						3	1		1
Procordulia affinis											2		1
Family Libellulidae													
Austrothemis nigrescens	4										2		
Diplacodes bipunctata	1			1						4			
Orthetrum caledonicum		1		1								2	1

COO   Coolgardie   12,912,209 ha

Granite strata of Yilgarn Craton with Archaean Greenstone intrusions in parallel belts. Drainage is occluded. Mallees and scrubs on sandplains associated with lateritised uplands, playas and granite outcrops. Diverse woodlands rich in endemic eucalypts, on low greenstone hills, valley alluvials and broad plains of calcareous earths. In the west, the scrubs are rich in endemic Proteaceae, in the east they are rich in endemic acacias. Arid to Semi-arid Warm Mediterranean.

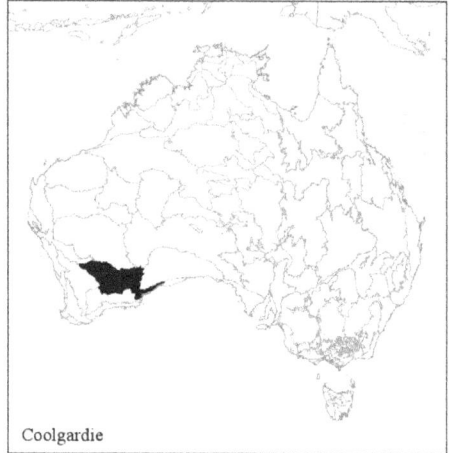

Coolgardie

Species   10   Specimens   52   Adults   52   Larvae   0

	J	F	M	A	M	J	J	A	S	O	N	D	n.d.
Family Lestidae													
Austrolestes annulosus	7											1	1
Austrolestes aridus										1			
Family Coenagrionidae													
Austroagrion cyane												3	
Ischnura aurora	1									1			
Xanthagrion erythroneurum									1			4	
Family Aeshnidae													
Hemianax papuensis			2										
Family Corduliidae													
Hemicordulia tau	1	1	1	1	1				2	1	1		
Family Libellulidae													
Diplacodes bipunctata	3								1			1	
Orthetrum caledonicum	8						1			1	1	1	
Pantala flavescens					1								

AVW   Avon Wheatbelt   9,517,104 ha

Area of active drainage dissecting a Tertiary plateau in Yilgarn Craton. Gently undulating landscape of low relief. Proteaceous scrub-heaths, rich in endemics, on residual lateritic uplands and derived sandplains; mixed eucalypt, *Allocasuarina huegeliana* and Jam-York Gum woodlands on Quaternary alluvials and eluvials. Semi-arid (Dry) Warm Mediterranean. The south eastern boundary has been modified incorporating a small portion into the Mallee region. Extensively cleared for agriculture.

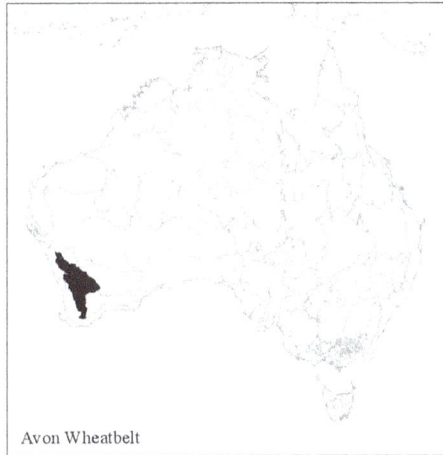

Avon Wheatbelt

Species   19   Specimens   154   Adults   154   Larvae   0

	J	F	M	A	M	J	J	A	S	O	N	D	n.d.
Family Lestidae													
Austrolestes analis			1			1				3			
Austrolestes annulosus	1	1	5	3					2	5	3	1	2
Austrolestes aridus								5	2	2			
Austrolestes io					1					5			
Family Argiolestidae													
Archiargiolestes pusillus											2	2	
Family Coenagrionidae													
Austroagrion cyane											1	2	2
Ischnura aurora			1	1									2
Ischnura heterosticta			4						1		5	3	3
Xanthagrion erythroneurum			4							3	1	2	2
Family Aeshnidae													
Hemianax papuensis	1	2	2					1				1	1
Family Gomphidae													
Austrogomphus collaris												8	2

	J	F	M	A	M	J	J	A	S	O	N	D	n.d.
Family Corduliidae													
Hemicordulia australiae											1		1
Hemicordulia tau	1	1	1		2			5		3	1		2
Family Libellulidae													
Austrothemis nigrescens											1		
Diplacodes bipunctata		1	1	1	1	3			1		2	1	1
Diplacodes haematodes			1										2
Orthetrum caledonicum	5	2	1								3	2	2
Pantala flavescens			1										
Tramea stenoloba		1											

JAF   Jarrah Forest   4,509,074 ha

Duricrusted plateau of Yilgarn Craton characterised by Jarrah-Marri forest on laterite gravels and, in the eastern part, by Marri-Wandoo woodlands on clayey soils. Eluvial and alluvial deposits support *Agonis* shrublands. In areas of Mesozoic sediments, Jarrah forests occur in a mosaic with a variety of species-rich shrublands. Warm Mediterranean climate.

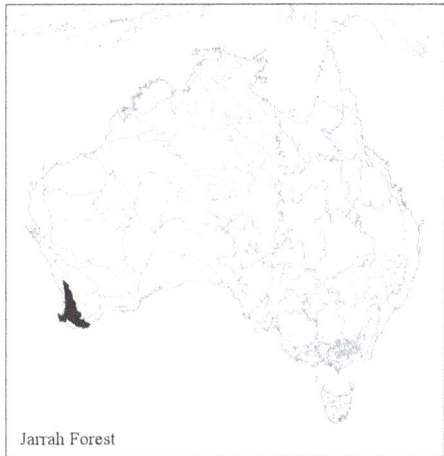

Jarrah Forest

Species   36   Specimens   768   Adults   768   Larvae   0

	J	F	M	A	M	J	J	A	S	O	N	D	n.d.
Family Lestidae													
Austrolestes aleison		10		1							2		
Austrolestes analis	9		1						1	16	7	2	
Austrolestes annulosus	2								1	1	4	10	
Austrolestes aridus										1			
Austrolestes io	3	1	2				1		2	3	2	4	
Family Argiolestidae													
Archiargiolestes parvulus												4	
Archiargiolestes pusillissimus										1	4	2	
Miniargiolestes minimus	14	2	1							2	6	22	1
Family Coenagrionidae													
Austroagrion cyane				4						22	2	5	
Ischnura aurora	1		1								2	3	
Ischnura heterosticta												2	4
Xanthagrion erythroneurum										1	1	5	
Family Aeshnidae													
Adversaeschna brevistyla	7		2						1	5	8	5	1
Austroaeschna anacantha	5	4	1					2			2	16	3
Hemianax papuensis	1		1	2						2	1	4	

	J	F	M	A	M	J	J	A	S	O	N	D	n.d.
Family Petaluridae													
Petalura hesperia	2								1			7	
Family Gomphidae													
Armagomphus armiger				1						13	4	5	
Austrogomphus collaris	8									9	7	36	2
Austrogomphus pusillus	2							6		19	42	41	2
Zephyrogomphus lateralis	1										4	11	
Family Synthemistidae													
Archaeosynthemis leachii	4	3	1									10	
Archaeosynthemis occidentalis	7	1	1						1		9	20	
Archaeosynthemis spiniger	2	1									1		
Austrosynthemis·cyanitincta				1			1			3	7	18	
Family Corduliidae													
Hemicordulia australiae	6		1		7				1		3	9	
Hemicordulia tau	1	1	3	7	1				3	4	8	7	
Procordulia affinis	2									2	1		
Family Libellulidae													
Austrothemis nigrescens	5										2	9	
Diplacodes bipunctata	3						1	2	1	1	4		
Diplacodes haematodes	5		1						4	2	4	4	
Nannophya occidentalis								2	1	1	2		
Orthetrum caledonicum	7	1	2	3					1	11	10		
Pantala flavescens			1										
Tramea stenoloba											1		
Genera incertae sedis													
Hesperocordulia berthoudi	1			1						2	2	6	
Lathrocordulia metallica				1							2	3	

WAR   Warren   844,771 ha

Dissected undulating country
of the Leeuwin Complex and
Albany Orogen with loamy
soils supporting Karri forest,
laterites supporting Jarrah-
Marri forest, leached sandy
soils in depressions and
plains supporting paperbark/
sedge swamps, and Holocene
marine dunes with *Agonis
flexuosa* woodlands. Moderate
Mediterranean.

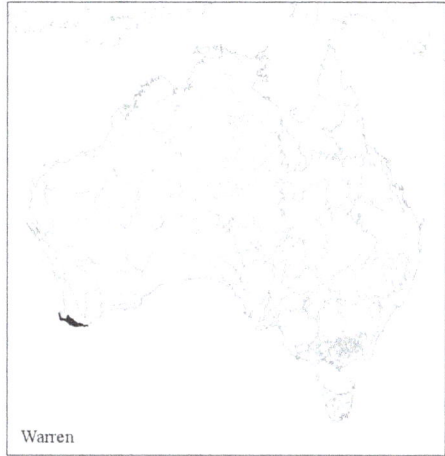

Warren

Species   28   Specimens   261   Adults   261   Larvae   0

	J	F	M	A	M	J	J	A	S	O	N	D	n.d.
Family Lestidae													
Austrolestes aleison			·								2	2	
Austrolestes analis		2		2							3	7	2
Austrolestes annulosus											1		
Austrolestes io	1	1									8	5	1
Family Argiolestidae													
Archiargiolestes parvulus											1		
Archiargiolestes pusillissimus	1	1								3	26	3	
Archiargiolestes pusillus										27	25	5	
Miniargiolestes minimus	13		2							3	8	6	
Family Coenagrionidae													
Austroagrion cyane										1	6	1	
Ischnura aurora				1							1		
Ischnura heterosticta											1		
Xanthagrion erythroneurum											1		
Family Aeshnidae													
Adversaeschna brevistyla	1											4	
Austroaeschna anacantha	3		1									5	
Hemianax papuensis											1	3	

	J	F	M	A	M	J	J	A	S	O	N	D	n.d.
Family Gomphidae													
Armagomphus armiger	1										1	1	
Austrogomphus collaris	3											6	
Zephyrogomphus lateralis	2										2	10	
Family Synthemistidae													
Archaeosynthemis leachii	2	1								1	2	5	
Archaeosynthemis occidentalis												1	
Austrosynthemis·cyanitincta											2	3	
Family Corduliidae													
Hemicordulia tau	1									2	4		
Procordulia affinis										3		1	
Family Libellulidae													
Diplacodes bipunctata	1										1	2	1
Diplacodes haematodes				1							1	2	
Orthetrum caledonicum		2										3	
Genera incertae sedis													
Hesperocordulia berthoudi												2	
Lathrocordulia metallica												1	

SWA   Swan Coastal Plain   1,525,798 ha

Low lying coastal plain, mainly covered with woodlands. It is dominated by Banksia or Tuart on sandy soils, *Casuarina obesa* on outwash plains, and paperbark in swampy areas. In the east, the plain rises to duricrusted Mesozoic sediments dominated by Jarrah woodland. Warm Mediterranean. Three phases of marine sand dune development provide relief. The outwash plains, once dominated by *Casuarina obesa*-marri woodlands and *Melaleuca* shrublands, are extensive only in the south.

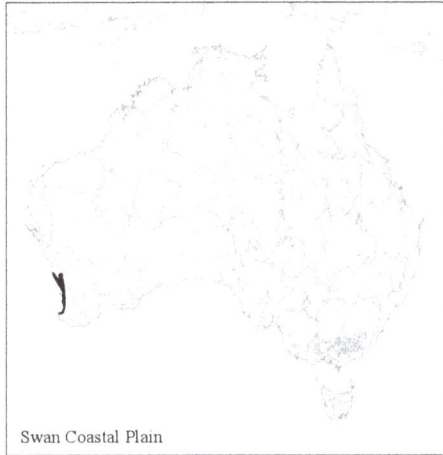

Swan Coastal Plain

Species   32   Specimens   1272   Adults   1272   Larvae   0

	J	F	M	A	M	J	J	A	S	O	N	D	n.d.
Family Lestidae													
Austrolestes aleison								2	1	2	2	7	
Austrolestes analis	7	5	7	4		1			7	25	10	16	1
Austrolestes annulosus	7	2		2	2	1	1		7	6	15	10	3
Austrolestes aridus									1	2			
Austrolestes io	3		1	2		3	1	3	3	4	4	7	
Family Argiolestidae													
Archiargiolestes parvulus										8	30	8	
Archiargiolestes pusillus	1									32	28	21	
Miniargiolestes minimus	5	1		1							14	9	
Family Coenagrionidae													
Austroagrion cyane	1								1	7	5	5	4
Ischnura aurora	5	5	1			1				10	1	7	
Xanthagrion erythroneurum	7	2	2	3					2	4	2	15	4
Family Aeshnidae													
Adversaeschna brevistyla	10	5	3	3	3	2		1	3	8	10	7	1
Austroaeschna anacantha	12	2	2		2						6	3	
Austrogynacantha heterogena	1												

	J	F	M	A	M	J	J	A	S	O	N	D	n.d.
Hemianax papuensis	10	7	5	9	2	10	1			2	6	13	2
Family Petaluridae													
Petalura hesperia	4										1	30	
Family Gomphidae													
Armagomphus armiger	2										5	1	
Austroepigomphus praeruptus													
Austroepigomphus gordoni								1					1
Austrogomphus collaris	6	2			1					24	26	14	9
Zephyrogomphus lateralis	4										2	5	
Family Synthemistidae													
Archaeosynthemis leachii	7										3	8	
Archaeosynthemis occidentalis	2									1	10	7	
Austrosynthemis·cyanitincta	1									1	6	4	1
Family Corduliidae													
Hemicordulia australiae	2	1	6	1	1					8	7	4	
Hemicordulia tau	7	4	2	7	12	11		1	11	17	18	14	1
Procordulia affinis	1								7	7	9		1
Family Libellulidae													
Austrothemis nigrescens	12	3	2	1			1		8	23	22		1
Diplacodes bipunctata	10	12	3	2	1			1	4	14	14	22	1
Diplacodes haematodes	3	32	2	1					1	6	6	3	8
Nannophya occidentalis			1							1	1		
Orthetrum caledonicum	6	7	8						3	5	21	32	9
Pantala flavescens			6										
Tramea loewii			1										
Tramea stenoloba	1	1	6				1			1	6	4	2
Genera incertae sedis													
Hesperocordulia berthoudi	3	1	1							1	13	2	
Lathrocordulia metallica												2	

GES   Geraldton Sandplains   3,142,149 ha

Mainly proteaceous scrub-
heaths, rich in endemics,
on the sandy earths of an
extensive, undulating, lateritic
sandplain mantling Permian to
Cretaceous strata. Extensive
York Gum and Jam woodlands
occur on outwash plains
associated drainage. Semi-arid
(Dry) warm Mediterranean.

Geraldton Sandplains

Species   22   Specimens   74   Adults   74   Larvae   0

	J	F	M	A	M	J	J	A	S	O	N	D	n.d.
Family Lestidae													
Austrolestes analis						3							
Austrolestes annulosus										1	1		
Family Argiolestidae													
Miniargiolestes minimus												1	
Family Coenagrionidae													
Austroagrion cyane									5				
Ischnura aurora									1				
Ischnura heterosticta			1						6		3		
Xanthagrion erythroneurum									2		2		
Family Aeshnidae													
Adversaeschna brevistyla									2				
Hemianax papuensis				1		1			2		2		
Family Gomphidae													
Austroepigomphus gordoni	1										1		
Austrogomphus collaris													1
Austrogomphus pusillus									2				
Ictinogomphus dobsoni	1												
Family Corduliidae													
Hemicordulia australiae									1				

	J	F	M	A	M	J	J	A	S	O	N	D	n.d.
Hemicordulia tau				2		1	1		3	2			
Family Libellulidae													
Austrothemis nigrescens											1		
Crocothemis nigrifrons											1		
Diplacodes bipunctata			1		1				2		1		
Diplacodes haematodes									2	1	2		
Macrodiplax cora										2	1		
Orthetrum caledonicum				2							3	2	
Zyxomma elgneri											1		

YAL   Yalgoo   5,087,577 ha

This region is an interzone between South-western Bioregions and Murchison. It is characterised by low woodlands to open woodlands of *Eucalyptus*, *Acacia* and *Callitris* on red sandy plains of the Western Yilgarn Craton and southern Carnarvon Basin. The latter has a basement of Phanerozoic sediments. This Bioregion extends westwards to the boundary of the South-west Botanical Province, so that it includes the Toolonga Plateau of the southern

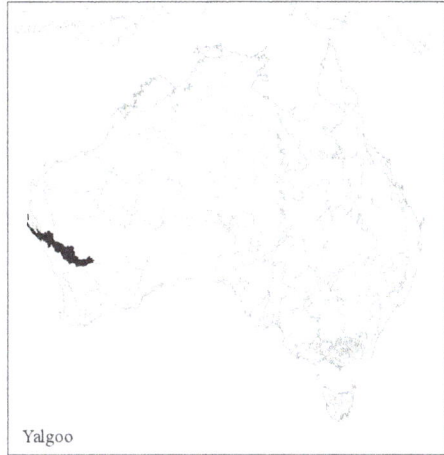
Yalgoo

Carnarvon Basin. Semi-arid to arid, warm, Mediterranean climate. Mulga, Callitris-*Eucalyptus salubris*, and Bowgada open woodlands and scrubs on earth to sandy-earth plains in the western Yilgarn Craton. Rich in ephemerals. Arid to semi-arid warm Mediterranean.

Species   11   Specimens   32   Adults   32   Larvae   0
This represents only 7 collecting events

	J	F	M	A	M	J	J	A	S	O	N	D	n.d.
Family Lestidae													
Austrolestes annulosus	1												
Family Coenagrionidae													
Ischnura aurora										3			
Xanthagrion erythroneurum										1			
Family Aeshnidae													
Hemianax papuensis	2									4			
Family Corduliidae													
Hemicordulia australiae									1				
Hemicordulia tau										6			
Family Libellulidae													
Crocothemis nigrifrons									4				

	J	F	M	A	M	J	J	A	S	O	N	D	n.d.
Diplacodes bipunctata			2						1				
Diplacodes haematodes									1	1			
Macrodiplax cora									2				
Orthetrum caledonicum									1	1			

MUR   Murchison   28,120,554 ha

Mulga low woodlands, often rich in ephemerals, on outcrop hardpan wash plains and fine-textured Quaternary alluvial and eluvial surfaces mantling granitic and greenstone strata of the northern part of the Yilgarn Craton. Surfaces associated with the occluded drainage occur throughout with hummock grasslands on Quaternary sandplains, saltbush shrublands on calcareous soils and *Halosarcia* low shrublands on saline alluvia. Areas of red sandplains with mallee-mulga parkland over hummock grasslands occur in the east.

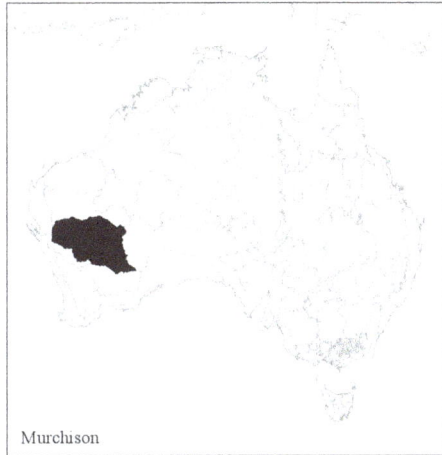

Murchison

Species   13   Specimens   68   Adults   68   Larvae   0

	J	F	M	A	M	J	J	A	S	O	N	D	n.d.
Family Lestidae													
Austrolestes annulosus						1							
Austrolestes aridus					4								
Family Coenagrionidae													
Agriocnemis argentea								1					
Austroagrion cyane			1										
Ischnura aurora	1		1										
Xanthagrion erythroneurum					3	1							
Family Aeshnidae													
Hemianax papuensis		1	3		3								2
Family Gomphidae													
Austroepigomphus gordoni								2					
Family Corduliidae													
Hemicordulia tau			4		7								
Family Libellulidae													
Diplacodes bipunctata			2	6	4	4				4			

	J	F	M	A	M	J	J	A	S	O	N	D	n.d.
Orthetrum caledonicum				2	3								1
Pantala flavescens		1											
Tramea loewii					4			1					

CAR   Carnarvon   8,430,172 ha

Quaternary alluvial, aeolian
and marine sediments
overlying Cretaceous strata. A
mosaic of saline alluvial plains
with samphire and saltbush
low shrublands, Bowgada low
woodland on sandy ridges
and plains, Snakewood scrubs
on clay flats, and tree to
shrub steppe over hummock
grasslands on and between red
sand dune fields. Limestone
strata with *Acacia startii /
bivenosa* shrublands outcrop
in the north, where extensive
tidal flats in sheltered
embayments support Mangal. Arid

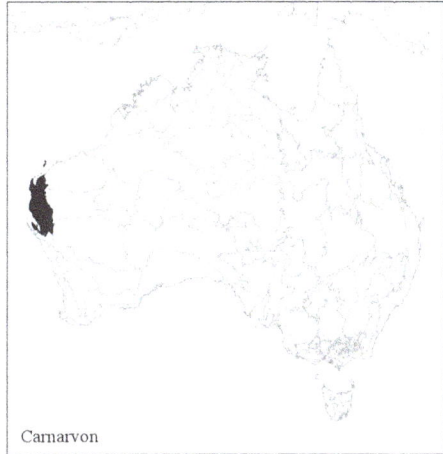

Carnarvon

Species   14   Specimens   85   Adults   85   Larvae   0

	J	F	M	A	M	J	J	A	S	O	N	D	n.d.
Family Lestidae													
Austrolestes aridus				1									
Austrolestes io				1									
Family Coenagrionidae													
Ischnura aurora					1								
Pseudagrion aureofrons													1
Xanthagrion erythroneurum					4					4			
Family Aeshnidae													
Hemianax papuensis		1	1	6	6			1		1			
Family Gomphidae													
Austroepigomphus gordoni	1										2	2	
Ictinogomphus dobsoni												1	
Family Corduliidae													
Hemicordulia intermedia			1										
Family Libellulidae													
Diplacodes bipunctata			1	4	12		1						
Diplacodes haematodes	3							2			1		1

	J	F	M	A	M	J	J	A	S	O	N	D	n.d.
Macrodiplax cora					3								
Pantala flavescens			1	4	11								
Tramea loewii					4								

PIL   Pilbara   17,823,126 ha

There are four major components to the Pilbara Craton. (1) Hamersley. Mountainous area of Proterozoic sedimentary ranges and plateaux with Mulga low woodland over bunch grasses on fine textured soils and Snappy Gum over *Triodia brizoides* on skeletal sandy soils of the ranges. (2) The Fortescue Plains. Alluvial plains and river frontages. Salt marsh, mulga-bunch grass, and short grass communities on alluvial plains. River Gum woodlands fringe the drainage lines. This is the northern limit of Mulga (*Acacia aneura*). (3) Chichester. Archaean granite and basalt plains supporting shrub steppe characterised by *Acacia pyrifolia* over *Triodia pungens* hummock grasses. Snappy Gum tree steppes occur on ranges. (4) Roebourne. Quaternary alluvial plains with a grass savanna of mixed bunch and hummock grasses, and dwarf shrub steppe of *Acacia translucens* over *Triodia pungens*. Samphire, *Sporobolus* and Mangal occur on marine alluvial flats. Arid tropical with summer rain.

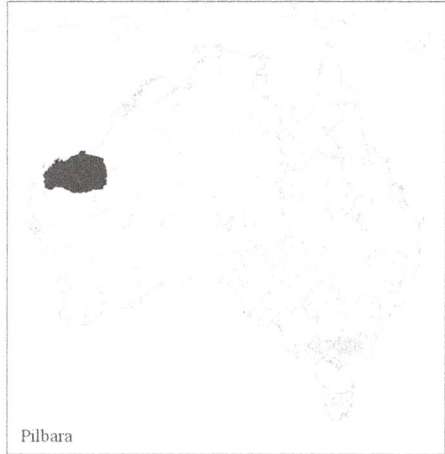

Pilbara

Species   37   Specimens   595   Adults   595   Larvae   0

	J	F	M	A	M	J	J	A	S	O	N	D	n.d.
Family Lestidae													
Austrolestes aridus	1	1											
Family Isostictidae													
Eurysticta coolawanyah		8								1	3	3	
Family Platycnemidae													
Nososticta liveringa		1									2		
Nososticta pilbara	8	2					1			2	9	2	
Family Coenagrionidae													
Agriocnemis argentea	2	2		4	1		5	2		2	9	1	
Agriocnemis kunjina	1	6					3	2		1	16	5	2

	J	F	M	A	M	J	J	A	S	O	N	D	n.d.
Argiocnemis rubescens										2			
Austroagrion pindrina	2						7	11			9	4	
Austrocnemis maccullochi										2			
Ischnura aurora	2	3		2			1				5	5	1
Ischnura heterosticta	2	4		4			2				4		
Pseudagrion aureofrons		2					5	16			5	2	
Pseudagrion microcephalum	6	17		2	1		4			8	10	2	
Xanthagrion erythroneurum		4		1							2	2	
Family Aeshnidae													
Adversaeschna brevistyla	1			1									
Austrogynacantha heterogena			1										
Hemianax papuensis	4			2	3	5		2			1	1	2
Family Gomphidae													
Antipodogomphus hodgkini	5	7		1									1
Austroepigomphus gordoni	1	2								1	8		
Austrogomphus mjobergi	2	6		3				2		8	4		1
Ictinogomphus dobsoni	8	7		1	1					1	9	4	
Family Corduliidae													
Hemicordulia intermedia				1	1								
Hemicordulia koomina	1							4			3	1	
Hemicordulia tau	5	1		1	1			2					
Family Libellulidae													
Crocothemis nigrifrons	4	3						2		1	3	1	
Diplacodes bipunctata	3	1	1	2	5	2	2				1	5	
Diplacodes haematodes	5	3	2	3	6	4	4		2	2	8	5	1
Macrodiplax cora												1	
Nannophlebia injibandi	4	6									3		2
Orthetrum caledonicum	8	4	2		6		2	1			4	1	
Orthetrum migratum	5	2	1	2	1		2			1	12	1	
Pantala flavescens	1	2	4		1								
Rhodothemis lieftincki	3	2					1				4	3	
Rhyothemis graphiptera	3	2								1	4		
Tramea loewii				1	1						2		
Tramea stenoloba	1	2									1	1	
Zyxomma elgneri		2			1					2	2		

DAL   Dampierland   8,360,871 ha

(1) Quaternary sandplain overlying Jurassic and Mesozoic sandstones with Pindan. Hummock grasslands on hills.

(2) Quaternary marine deposits on coastal plains, with Mangal, samphire - *Sporobolus* grasslands, *Melaleuca acacioides* low forests, and *Spinifex - Crotalaria* strand communities.

(3) Quaternary alluvial plains associated with the Permian and Mesozoic sediments of Fitzroy Trough support tree

Dampierland

savannas of *Chrysopogon - Dichanthium* grasses with scattered *Eucalyptus microtheca - Lysiphyllum cunninghamii*. Riparian forests of River Gum and Cadjeput fringe drainages.

(4) Devonian reef limestones in the north and east support sparse tree steppe over *Triodia intermedia* and *T. wiseana* hummock grasses and vine thicket elements.

Dry hot tropical, semi-arid summer rainfall.

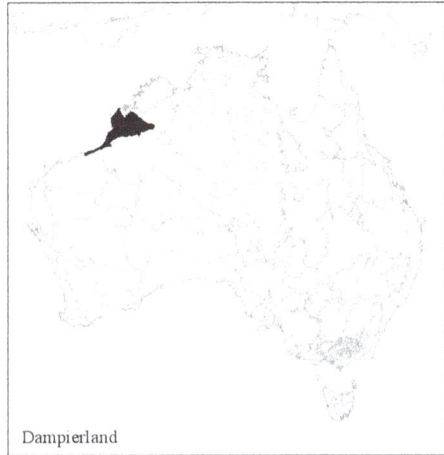

Species   23   Specimens   155   Adults   155   Larvae   0

	J	F	M	A	M	J	J	A	S	O	N	D	n.d.
Family Lestidae													
Austrolestes annulosus	1												
Austrolestes insularis								1					
Lestes concinnus		2						1					
Family Isostictidae													
Austrosticta soror			11										
Rhadinosticta banksi		3	9										
Family Platycnemidae													
Nososticta liveringa		18						4					
Family Coenagrionidae													
Agriocnemis pygmaea			4							1			
Ceriagrion aeruginosum										1			

	J	F	M	A	M	J	J	A	S	O	N	D	n.d.
Ischnura aurora		1	2										
Pseudagrion aureofrons		8	14					3					
Pseudagrion microcephalum		16	11					2					
Family Aeshnidae													
Austrogynacantha heterogena		1											
Hemianax papuensis			1		2								
Family Gomphidae													
Austrogomphus mjobergi		1											
Austrogomphus pusillus													1
Family Libellulidae													
Diplacodes bipunctata		2	3				1	1					
Diplacodes haematodes		1	5	1			2						
Neurothemis stigmatizans								1					
Orthetrum caledonicum	1		7					1					
Orthetrum sabina			2										
Pantala flavescens			3		2								
Potamarcha congener								1					
Tramea loewii								2					

188

CEK   Central Kimberley   7,675,587 ha

Hilly to mountainous country with parallel siliceous ranges of Proterozoic sedimentary rocks with skeletal sandy soils supporting *Plectrachne pungens* hummock grasses with scattered trees, and with earths on Proterozoic volcanics in valleys supporting Ribbon Grass with scattered trees. Open forests of River Gum and Pandanus occur along drainage lines. Dry hot tropical, sub-humid to semi-arid, summer rainfall.

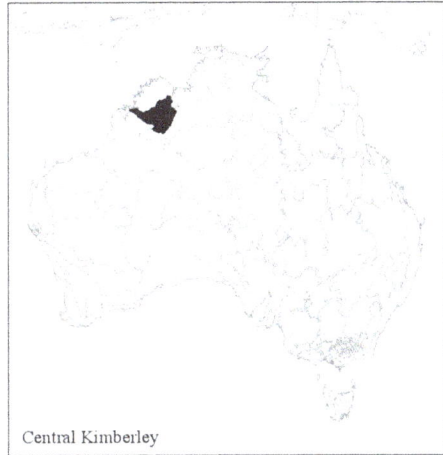

Central Kimberley

Species   25   Specimens   106   Adults   106   Larvae   0

	J	F	M	A	M	J	J	A	S	O	N	D	n.d.
Family Lestidae													
Austrolestes insularis							1	2					
Family Platycnemidae													
Nososticta kalumburu							3	3			1		
Nososticta liveringa							2	23					
Family Coenagrionidae													
Argiocnemis rubescens							1	4					
Austroagrion exclamationis							1	2					
Austroagrion watsoni							1	2	1				
Austrocnemis obscura								1					
Ischnura aurora								1					
Pseudagrion aureofrons								1					
Pseudagrion lucifer							1	3					
Pseudagrion microcephalum								1					
Xanthagrion erythroneurum											1		
Family Aeshnidae													
Gynacantha nourlangie						2	4	2	1		1		
Family Gomphidae													
Austroepigomphus turneri				1									

	J	F	M	A	M	J	J	A	S	O	N	D	n.d.
Austrogomphus mouldsorum												1	
Ictinogomphus australis												1	
Family Libellulidae													
Diplacodes bipunctata				4									
Diplacodes haematodes				9			3						
Nannodiplax rubra							2						
Neurothemis stigmatizans							4						
Orthetrum caledonicum				3			1						
Orthetrum migratum							4						
Pantala flavescens				1									1
Tramea loewii							1	1					
Tramea stenoloba				1									

NOK   Northern Kimberley   8,420,100 ha

Dissected plateau of Kimberley Basin. Savanna woodland of Woolybutt and Darwin Stringy bark over high Sorghum grasses and *Plectrachne schinzii* hummock grasses on shallow sandy soils on outcropping Proterozoic siliceous sandstone strata. Savanna woodlands on *Eucalyptus tectifica - E. grandiflora* alliance over high Sorghum grasses on red and yellow earths mantling basic Proterozoic volcanics. Riparian closed forests of paperbark trees and *Pandanus* occur along drainage lines. Extensive Mangal occurs in estuaries and sheltered embayments. Numerous small patches of monsoon rainforest are scattered through the district. Dry hot tropical, sub-humid, summer rainfall.

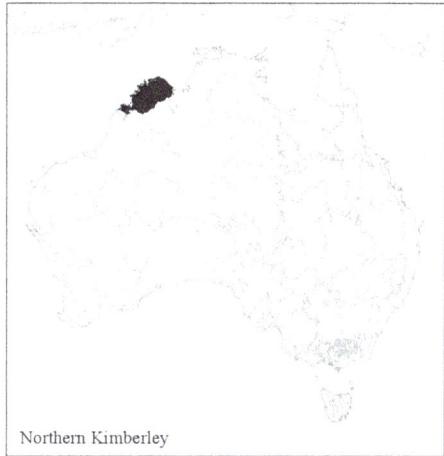

Northern Kimberley

Species   44   Specimens   311   Adults   311   Larvae   0

	J	F	M	A	M	J	J	A	S	O	N	D	n.d.
Family Lestidae													
Austrolestes insularis					1	1							
Family Isostictidae													
Austrosticta fieldi					1	1							
Austrosticta soror			1										
Family Platycnemidae													
Nososticta kalumburu		1					1	19					
Nososticta liveringa								56					
Family Coenagrionidae													
Agriocnemis pygmaea								1		1			
Argiocnemis rubescens							1	4					
Austroagrion exclamationis							1	2					
Austroagrion watsoni							1	5					
Austrocnemis obscura							1						1
Ceriagrion aeruginosum			1				1	1					

	J	F	M	A	M	J	J	A	S	O	N	D	n.d.
Ischnura aurora								3					
Ischnura heterosticta					1					1			
Pseudagrion lucifer						1	2	4					
Pseudagrion microcephalum							4	3					
Family Aeshnidae													
Anax georgius								1					
Austrogynacantha heterogena		2		1	3							1	
Gynacantha dobsoni								3					
Gynacantha nourlangie					5	2		10	2	4			
Hemianax papuensis		6	1	1								1	
Family Gomphidae													
Antipodogomphus neophytus		1											
Austroepigomphus turneri									4				
Ictinogomphus australis		1											
Family Corduliidae													
Hemicordulia intermedia		1						1					
Family Libellulidae													
Aethriamanta circumsignata		1											
Brachydiplax denticauda		1											
Diplacodes bipunctata	1	4			2			3	1				
Diplacodes haematodes	1	2			6	4	1	5	4				
Macrodiplax cora		1	1		1			1					
Nannodiplax rubra		1			1	2		8		1			
Nannophlebia injibandi							1						
Nannophlebia mudginberri								1					
Neurothemis stigmatizans		1		1	2	1		4	3	1		1	
Notolibellula bicolor						2							
Orthetrum caledonicum	2	3											
Orthetrum migratum						3		4	4				
Orthetrum sabina		1											
Pantala flavescens	3	1	1	1		1			1				
Rhyothemis braganza		1	1					2	1				
Rhyothemis graphiptera			6							3			
Tholymis tillarga	2	1									2		
Tramea loewii	1	4			2	2		3	2	1		3	
Tramea stenoloba					1	1							
Zyxomma elgneri	1	2											

VIB   Victoria Bonaparte   7,301,242 ha

Phanerozoic strata of the
Bonaparte Basin in the north-
western part are mantled by
Quaternary marine sediments
supporting Samphire -
Sporobolus grasslands and
mangal, and by red earth
plains and black soil plains
with an open savanna of
high grasses. Outcrops of
Devonian limestone karst in
the west support tree steppe
and vine thicket. Plateaux and
abrupt ranges of Proterozoic
sandstone, known as the
Victoria Plateau, occur in the
south and east, and are partially mantled by skeletal sandy soils
with low tree savannas and hummock grasslands. In the south
east are limited areas of gently undulating terrain on a variety of
sedimentary rocks supporting low Snappy Gum over hummock
grasslands and also of gently sloping floodplains supporting
Melaleuca minutifolia low woodland over annual sorghums. Dry
hot tropical, semi-arid summer rainfall.

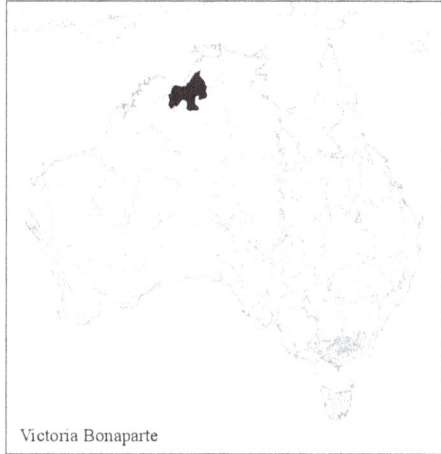

Victoria Bonaparte

| Species | 67 | Specimens | 375 | Adults | 375 | Larvae | 0 |

	J	F	M	A	M	J	J	A	S	O	N	D	n.d.
Family Lestidae													
Austrolestes insularis				1									
Indolestes alleni							1		1				
Lestes concinnus	2	3				1		1	1				
Family Isostictidae													
Austrosticta soror		1											
Eurysticta kununurra	1	2											
Rhadinosticta banksi		1											
Family Platycnemidae													
Nososticta baroalba							1		1	1			
Nososticta coelestina					2					1			
Nososticta fraterna									2	1			

	J	F	M	A	M	J	J	A	S	O	N	D	n.d.
Nososticta kalumburu	2												
Nososticta liveringa		38					4						
Family Coenagrionidae													
Aciagrion fragile								2					
Agriocnemis argentea								1					
Agriocnemis pygmaea							1	2	4				
Argiocnemis rubescens								1		1		1	
Austroagrion exclamationis				2				1	3				
Austroagrion watsoni								1					
Austrocnemis maccullochi								1					
Ceriagrion aeruginosum								1		2			
Ischnura aurora		1	1	2	1			2	5				
Ischnura heterosticta					1		1	2	3				
Ischnura pruinescens	1			1				1					
Pseudagrion aureofrons		3						1					
Pseudagrion cingillum		13					2	4	1				
Pseudagrion jedda					6								
Pseudagrion lucifer	1												
Pseudagrion microcephalum		4			1		1	2	5				
Xanthagrion erythroneurum		1						2					
Family Aeshnidae													
Anax gibbosulus									1				
Anax guttatus	2												
Austrogynacantha heterogena	1	8											
Gynacantha dobsoni							2						
Gynacantha nourlangie					1			1				1	
Hemianax papuensis	10	3	1									1	1
Family Gomphidae													
Antipodogomphus neophytus	2	1											
Austrogomphus mjobergi		3										1	
Austrogomphus mouldsorum												1	
Ictinogomphus australis					1			1					
Family Corduliidae													
Hemicordulia intermedia		1				1			1				
Pentathemis membranulata		1											
Family Libellulidae													
Brachydiplax denticauda					1								
Crocothemis nigrifrons	1	3						1		1			

	J	F	M	A	M	J	J	A	S	O	N	D	n.d.
Diplacodes bipunctata	4	5	2	5	1			1	2				2
Diplacodes haematodes	2	1	1	2			2	1	2	5			
Diplacodes nebulosa	1							2	1				
Diplacodes trivialis	2				1		1	1					1
Hydrobasileus brevistylus	3								2				
Lathrecista asiatica festa									1				
Macrodiplax cora		2			1							1	
Nannodiplax rubra	3	1	1	4	1	1		1					
Nannophlebia eludens										1			
Nannophlebia injibandi										1			
Neurothemis stigmatizans	3	1		1									
Notolibellula bicolor	1	1											
Orthetrum caledonicum	3	3		3					2	4			1
Orthetrum migratum		2							1	1			
Orthetrum sabina		1						5					
Orthetrum serapia					1								
Pantala flavescens		5	1		1	1			1				1
Potamarcha congener		4											
Rhodothemis lieftincki	1	1											
Rhyothemis braganza	3	1	1	2						1			
Rhyothemis graphiptera	3		2	2	1		1	1					
Tholymis tillarga		2			1		1						
Tramea loewii	2	6			2		1						
Urothemis aliena								1					
Zyxomma elgneri			1		1				1				

195

OVP   Ord Victoria Plain   12,540,703 ha

Level to gently undulating plains with scattered hills on Cambrian volcanics and Proterozoic sedimentary rocks; vertosols on plains and predominantly skeletal soils on hills; grassland with scattered Bloodwood and Snappy Gum with spinifex and annual grasses. Dry hot tropical, semi-arid summer rainfall. The lithological mosaic has three main components: (1) Abrupt Proterozoic and Phanerozoic ranges and scattered hills mantled by shallow sand and loam soils supporting *Triodia* hummock grasslands with sparse low trees. (2) Cambrian volcanics and limestones form extensive plains with short grass (*Enneapogon* spp.) on dry calcareous soils and medium-height grassland communities (*Astrebla* and *Dichanthium*) on cracking clays. Riparian forests of River Gums fringe drainage lines. (3) In the south-west, Phanerozoic strata expressed as often lateritised upland sandplains with sparse trees. This component recurs as the Sturt Plateau Region in central Northern Territory.

Ord Victoria Plain

Species   25   Specimens   73   Adults   73   Larvae   0

	J	F	M	A	M	J	J	A	S	O	N	D	n.d.
Family Isostictidae													
Austrosticta fieldi				1									
Eurysticta coomalie				1									
Family Platycnemidae													
Nososticta fraterna						1	1						
Nososticta liveringa					12		1	2	1		4		
Family Coenagrionidae													
Agriocnemis pygmaea		1				2							
Argiocnemis rubescens						1							
Austroagrion exclamationis											2		
Austroagrion watsoni							1						

	J	F	M	A	M	J	J	A	S	O	N	D	n.d.
Ceriagrion aeruginosum							5						
Ischnura aurora							1						
Ischnura heterosticta						2							
Pseudagrion jedda						1	1						
Pseudagrion microcephalum							1						
Xanthagrion erythroneurum											2		
Family Aeshnidae													
Gynacantha nourlangie											2		
Family Gomphidae													
Austrogomphus mjobergi				1	1								
Family Libellulidae													
Diplacodes bipunctata				1	2	1							
Diplacodes haematodes					1	6				1	2		
Diplacodes trivialis						1					1		
Lathrecista asiatica festa											1		
Orthetrum caledonicum						3							
Pantala flavescens						1							
Rhyothemis graphiptera											2		
Tramea loewii									1				
Tramea stenoloba		1											

197

DAB   Daly Basin   2,092,229 ha

Gently undulating plains and
scattered low plateau remnants
on Palaeozoic sandstones,
siltstones and limestones;
neutral loamy and sandy red
earths; Darwin Stringybark
and Darwin Woollybutt open
forest with perennial and
annual grass understorey.

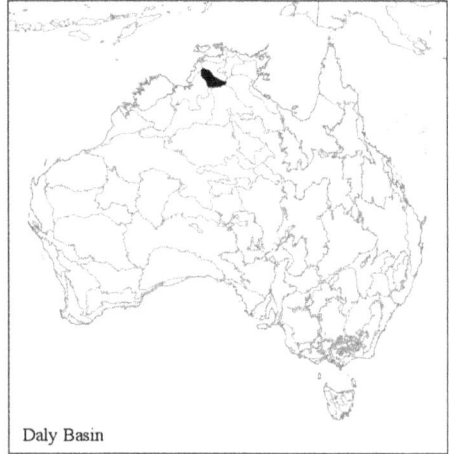

Daly Basin

Species   53   Specimens   254   Adults   254   Larvae   0

	J	F	M	A	M	J	J	A	S	O	N	D	n.d.
Family Lestidae													
Austrolestes insularis		2											
Lestes concinnus		5		1							2		
Family Isostictidae													
Austrosticta fieldi		1											
Eurysticta coomalie		1									1		
Rhadinosticta banksi		1											
Family Platycnemidae													
Nososticta coelestina					4					1	2		
Nososticta fraterna		1		1	7	1		3	2	1	7		
Family Coenagrionidae													
Aciagrion fragile							1						
Agriocnemis pygmaea					1			1					
Argiocnemis rubescens					3		1						
Austroagrion exclamationis		1											
Ceriagrion aeruginosum					1								
Ischnura aurora		1			2	2							
Ischnura heterosticta		1				3							
Pseudagrion jedda		12			19		1	1			11		

	J	F	M	A	M	J	J	A	S	O	N	D	n.d.
Pseudagrion lucifer					3								
Pseudagrion microcephalum					7	6		1			2		
Family Aeshnidae													
Anax gibbosulus											1		
Austrogynacantha heterogena		1											
Gynacantha nourlangie						8	1						
Hemianax papuensis		1											
Family Gomphidae													
Antipodogomphus dentosus		4											
Antipodogomphus neophytus		3											
Austrogomphus mjobergi	1	17											
Ictinogomphus australis										1			
Family Macromiidae													
Macromia tillyardi	1												
Family Corduliidae													
Hemicordulia intermedia		1									2		
Pentathemis membranulata		1											
Family Libellulidae													
Agrionoptera i.allogenes					1		1						
Crocothemis nigrifrons											1		
Diplacodes bipunctata		3		1		1		1					
Diplacodes haematodes		1		5		1	1				1		
Diplacodes nebulosa		1				1							
Diplacodes trivialis											1		
Hydrobasileus brevistylus			1	1									
Macrodiplax cora											1		
Nannodiplax rubra		2			1		2	1			1		
Nannophlebia eludens					3								
Nannophlebia injibandi		1		5	3			1					
Nannophlebia mudginberri					2					2			
Neurothemis stigmatizans		1		4							1		
Orthetrum caledonicum		2		1	1			1					
Orthetrum migratum					2					1			
Orthetrum sabina						1							
Orthetrum villosovittatum					1								
Pantala flavescens		1				1							
Potamarcha congener		2											
Rhodothemis lieftincki					1		1			1	1		

	J	F	M	A	M	J	J	A	S	O	N	D	n.d.
Rhyothemis braganza											1		
Rhyothemis graphiptera		2		2		1	1	1					
Tholymis tillarga		1			1						1		
Tramea loewii				1		1							
Zyxomma elgneri		1			1								

DAC   Darwin Coastal   2,843,199 ha

Gently undulating plains on lateritised Cretaceous sandstones and siltstones; sandy and loamy red and yellow earths and siliceous sands from near the mouth of the Victoria River to just west of Cobourg Peninsula. The most notable vegetation feature is the extensive and diverse floodplain environment associated with the lower reaches of the many large river systems. There are also substantial areas of mangroves, and rainforest and other riparian vegetation fringing the rivers. Inland from the coast, the dominant vegetation type is eucalypt tall open forest, typically dominated by Darwin woollybutt (*Eucalyptus miniata*) and Darwin stringybark (*E. tetrodonta*). Large waterbird colonies are a major conservation value of the bioregion.

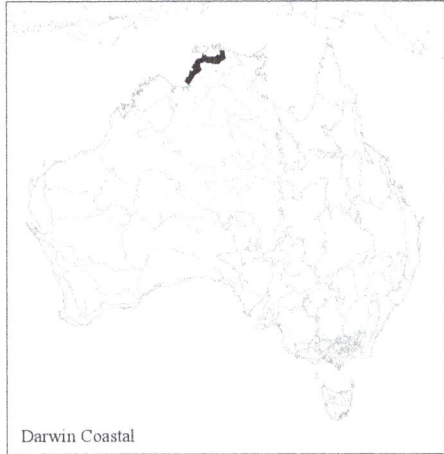

Darwin Coastal

Species   67   Specimens   885   Adults   885   Larvae   0

	J	F	M	A	M	J	J	A	S	O	N	D	n.d.
Family Lestidae													
Austrolestes insularis	1						1						
Indolestes alleni											1	1	
Indolestes obiri					5					2			
Family Argiolestidae													
Podopteryx selysi													1
Family Platycnemidae													
Nososticta baroalba							5		4				
Nososticta coelestina							1	2					
Nososticta fraterna	1				1		6					2	2
Nososticta koolpinyah	1		1		2			1		1	10		
Nososticta mouldsi						7						1	
Nososticta taracumbi	1								1				

	J	F	M	A	M	J	J	A	S	O	N	D	n.d.
Family Coenagrionidae													
Aciagrion fragile	1	1	1										
Agriocnemis pygmaea		1	1	5	1	3		3	1	2	4		
Archibasis mimetes							1						
Argiocnemis rubescens					1	1					1	1	
Austroagrion exclamationis	1	4	1		2	2	1	1			8		
Austroagrion watsoni								2					
Austrocnemis maccullochi		2					1	5					
Ceriagrion aeruginosum	2				2	2	2	10	5	5	4		1
Ischnura aurora					1	3		6			1		1
Ischnura heterosticta		1	4	12	6	8	1	8	4	7	3		
Ischnura pruinescens				5	4	1	3	6	1	1			2
Pseudagrion cingillum					1					7	1		
Pseudagrion jedda					1		16				1		
Pseudagrion lucifer							4	1			1	1	
Pseudagrion microcephalum		2	1		9	1	2	1			1		
Family Aeshnidae													
Anaciaeschna jaspidea	1		1	1									
Anax gibbosulus		2			1					2	2	1	
Anax guttatus	2	2	3	5	1			2	1		1		
Gynacantha dobsoni	1			1			1	1					
Gynacantha nourlangie					2		1	3		2			
Hemianax papuensis		1			1								2
Family Gomphidae													
Austroepigomphus turneri										2			
Ictinogomphus australis		1	1							2	4	2	
Family Corduliidae													
Hemicordulia kalliste	1												
Pentathemis membranulata												1	
Family Libellulidae													
Aethriamanta circumsignata	1				1								
Aethriamanta nymphaea			1					2					
Agrionoptera i.allogenes	1				1		1	1	2		3		
Brachydiplax denticauda			1	5	2		2	2	6	1	1	2	3
Camacinia othello		1											1
Crocothemis nigrifrons	1				2	4	1			1			1
Diplacodes bipunctata	2	3	5	5	1	2		4	1	1			2
Diplacodes haematodes						1		1					

	J	F	M	A	M	J	J	A	S	O	N	D	n.d.
Diplacodes nebulosa		1						2			1		
Diplacodes trivialis	4		6	3	4	5		1	7	9	4	1	2
Hydrobasileus brevistylus			1	12		2		4	3	2			2
Lathrecista asiatica festa	2	1	6		3	1		4	4		2	1	
Macrodiplax cora			1	1		1		2	5	1			
Nannodiplax rubra	2	2		2	1	3	1	6	8			1	1
Nannophlebia eludens		1						1			2		
Nannophlebia injibandi													1
Neurothemis stigmatizans	5	5	11	1	1	9	11	8	25	7	7	6	4
Orthetrum balteatum											1		
Orthetrum caledonicum	1		1		1			2	1	1			
Orthetrum migratum	1							1					
Orthetrum sabina	3		5	1	3	4		1	9	2	3	1	2
Orthetrum serapia						1							
Orthetrum villosovittatum	1				1						1		
Pantala flavescens	2	1	1	1	1	3					1	1	1
Potamarcha congener	1										1		
Rhodothemis lieftincki	3						3		2				
Rhyothemis braganza	1	1		1				2			4	1	2
Rhyothemis graphiptera	2	3	10	5	6	5	3	4	3	3	1		2
Rhyothemis phyllis	1	1	4					1		2	2	1	
Tholymis tillarga	1	3		4	4		2	2	1	1		1	
Tramea loewii	2		25	3	4	4	1	3	1	2	3		1
Tramea stenoloba									1	1			

TIW   Tiwi Cobourg   1,010,580 ha

This coastal region includes
Australia's second and fifth
largest islands (Melville and
Bathurst Island in the Tiwi
island group), Croker Island
and the adjacent Cobourg
Peninsula. Coastal vegetation
includes some mangroves
and saline flats, although
this bioregion lacks the
large rivers which influence
vegetation patterning in
other coastal regions. Most
of this bioregion is covered
by tall eucalypt open forests,
typically dominated by

Tiwi Cobourg

Darwin woollybutt (*Eucalyptus miniata*), Darwin stringybark (*E. tetrodonta*) and Melville Island bloodwood (*E. nesophila*), but often with northern cypress-pine *Callitris intratropica* and the tall palm *Gronophyllum ramsayi* co-dominant. The Tiwi Islands support a relatively high density and total area of monsoon rainforest patches, with distinctive species composition. There are also substantial areas there of a distinctive "treeless plain" vegetation. This bioregion is of low relief, with laterite and Cretaceous sandstone the dominant substrates. The Tiwi Islands support about 20 endemic plant and vertebrate animal taxa. The bioregions contains some important marine turtle breeding sites, and a Ramsar wetland on the Cobourg Peninsula. The bioregion is entirely Aboriginal land.

Species   44   Specimens 366   Adults   366   Larvae   0

	J	F	M	A	M	J	J	A	S	O	N	D	n.d.
Family Lestidae													
Austrolestes insularis		10											
Indolestes alleni						2							
Lestes concinnus							1						
Family Platycnemidae													
Nososticta fraterna										32			
Nososticta koolpinyah						13	1			60			

	J	F	M	A	M	J	J	A	S	O	N	D	n.d.
Nososticta taracumbi						10	1			21			
Family Coenagrionidae													
Aciagrion fragile										19			
Agriocnemis pygmaea										1			
Archibasis mimetes								1					
Argiocnemis rubescens										1			
Austroagrion exclamationis		3					1			12			
Austrocnemis maccullochi										5			
Ceriagrion aeruginosum								1		11			2
Ischnura heterosticta		5								3			
Pseudagrion lucifer						1							
Pseudagrion microcephalum										3			
Family Aeshnidae													
Anax gibbosulus							1	1		2			
Anax guttatus		1	2										
Austrogynacantha heterogena		1											
Family Gomphidae													
Austroepigomphus turneri										1			
Family Macromiidae													
Macromia tillyardi										1			
Family Corduliidae													
Pentathemis membranulata										4			
Family Libellulidae													
Aethriamanta circumsignata										1			
Agrionoptera i.allogenes		1				1							
Brachydiplax denticauda										1			
Diplacodes bipunctata	1	5								2			
Diplacodes haematodes							3	1		5			
Diplacodes trivialis	1	12								3			1
Huonia melvillensis										5			
Lathrecista asiatica festa		2								1			
Macrodiplax cora		2											
Nannodiplax rubra										16			1
Nannophlebia mudginberri										3			
Neurothemis stigmatizans		7					2	2		6	1		1
Orthetrum caledonicum							2			4			
Orthetrum migratum										5			
Orthetrum sabina		2								1			1

	J	F	M	A	M	J	J	A	S	O	N	D	n.d.
Orthetrum villosovittatum	1						1						
Pantala flavescens	1	7									1		
Rhyothemis braganza	1	2								2	2		
Tholymis tillarga		1						1				1	
Tramea loewii		5	1										
Tramea stenoloba		2											
Urothemis aliena										1			

PCK   Pine Creek   2,851,777 ha

Foothill environments below
and to the west of the western
Arnhem Land sandstone massif.
Its main defining feature is
the highly mineraliferous
Pine Creek Geosyncline,
comprising Archaean granite
and gneiss overlain by
Palaeoprotozoic sediments.
The major vegetation types
are eucalypt tall open forests,
typically dominated by Darwin
woollybutt (*Eucalyptus miniata*)
and Darwin stringybark (*E.
tetrodonta*), and woodlands
(dominated by a range of

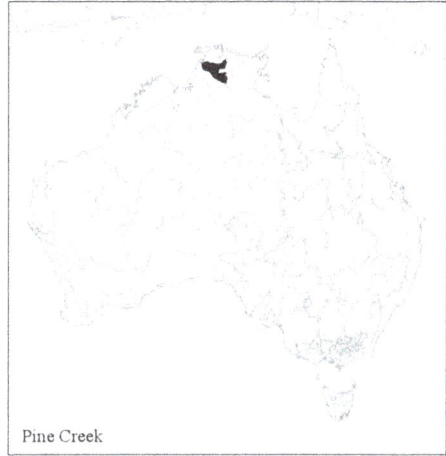

Pine Creek

species including *E. grandifolia*, *E. latifolia*, *E. tintinnans*, *E.
confertiflora* and *E. tectifica*), with smaller areas of monsoon
rainforest patches, *Melaleuca* woodlands, riparian vegetation and
tussock grasslands. Characteristic species include the granivorous
birds Gouldian finch *Erythrura gouldii*, hooded parrot *Psephotus
dissimilis* and partridge pigeon *Geophaps smithii*.

Species   74   Specimens 1048   Adults   1048   Larvae   0

	J	F	M	A	M	J	J	A	S	O	N	D	n.d.
Family Lestidae													
Austrolestes insularis	1				1		1			1	3		
Indolestes obiri			1		1						1		
Lestes concinnus	2	2	2		5		2				5		
Family Isostictidae													
Austrosticta fieldi			2		1						3		
Eurysticta coomalie		3	2										
Lithosticta macra			3	1	1								
Rhadinosticta banksi	1	1											
Family Platycnemidae													
Nososticta baroalba	1	7	11	2	10	1	3				13	7	
Nososticta coelestina							1	1				1	
Nososticta fraterna	3	1	2		14		5	4			20	9	

	J	F	M	A	M	J	J	A	S	O	N	D	n.d.
Nososticta koongarra			4		8	1		1			17		
Family Coenagrionidae													
Aciagrion fragile			1		4				1		15		
Agriocnemis pygmaea			1		6	1	1	1			11	1	
Agriocnemis rubricauda			3						1		1	1	
Archibasis mimetes											1		
Argiocnemis rubescens			4		7		2		1		7	2	
Austroagrion exclamationis			3		10	4	2	3		3	10	4	
Austroagrion watsoni											1		
Austrocnemis maccullochi					14	8		3			9	2	
Ceriagrion aeruginosum					2						1	5	
Ischnura aurora		1	1	2	4	1		1	4		3		
Ischnura heterosticta				1	11	8	1	3	1	2	9	3	
Ischnura pruinescens					4	1		4	3	1	4	2	
Pseudagrion cingillum					5	1					2		
Pseudagrion jedda	1				24		2				1		1
Pseudagrion lucifer					4								
Pseudagrion microcephalum			2		19	3	2				6	3	
Family Aeshnidae													
Anax gibbosulus			1		1				1				
Anax guttatus											1		
Austrogynacantha heterogena				3	2								
Gynacantha dobsoni			2								1	1	
Gynacantha nourlangie			2		6		4						1
Hemianax papuensis		1	1		1								
Family Gomphidae													
Antipodogomphus neophytus											1		
Austrogomphus mjobergi	1	1											
Hemigomphus magela				2							9		
Ictinogomphus australis					2						7		
Family Macromiidae													
Macromia tillyardi											1		
Family Corduliidae													
Hemicordulia intermedia			2	2			2				1	2	
Pentathemis membranulata											10	4	
Family Libellulidae													
Aethriamanta circumsignata			3		1						2		
Aethriamanta nymphaea						1		1			7		

	J	F	M	A	M	J	J	A	S	O	N	D	n.d.
Agrionoptera i.allogenes			1		3						2	1	
Brachydiplax denticauda			1	2	3					1	6	3	
Crocothemis nigrifrons			1		1					1	2	2	
Diplacodes bipunctata		1	7	2	6	2	1	1					
Diplacodes haematodes			2	7	3			3	2		3	2	
Diplacodes nebulosa		1	2		3						3	2	
Diplacodes trivialis			2		3	2		3		3	2	2	
Hydrobasileus brevistylus			3						1		5	2	
Lathrecista asiatica festa			2		1							1	
Macrodiplax cora			3		3				1		4	1	
Nannodiplax rubra	1		3	2	8	6	3	1	7		10	2	1
Nannophlebia eludens			1		4	1		1			5	1	
Nannophlebia injibandi				3	2								
Nannophlebia mudginberri			4		3	1	2				3		
Neurothemis stigmatizans	1	3	6	2	10	4	2	4	5	2	5	11	
Notolibellula bicolor				2									
Orthetrum balteatum											1	3	
Orthetrum caledonicum					2						3		
Orthetrum migratum			4	1	2		1	1			3	3	
Orthetrum sabina			5		3	1	1	2			4	1	
Orthetrum villosovittatum			1		2								
Pantala flavescens			3		2	2							
Potamarcha congener	1		2		2						3		
Rhodothemis lieftincki			1		1				2		7	2	
Rhyothemis braganza			4						1		3	9	
Rhyothemis graphiptera	2	3	5	2	4	1	2	3	6		4	3	
Rhyothemis phyllis											2	2	
Tholymis tillarga		2	4	1	5		1	1			5	1	
Tramea loewii		1	4	2	3	1	2	1			6	1	
Urothemis aliena					2						1	3	
Zyxomma elgneri		1			5			1		1	3	3	
Genera incertae sedis													
Austrocordulia territoria					9								

ARP   Arnhem Plateau   2,306,023 ha

The extensive and highly
dissected Proterozoic sandstone
massif of western Arnhem Land,
which forms the headwaters of
many of the major river systems
of the Top End. It supports
an unusually diverse biota,
including very many relictual
and endemic plant and animal
species. The major vegetation
types include sandstone
heathlands, rainforests
(characteristically dominated by
the endemic tree *Allosyncarpia
ternata*), hummock grasslands
and eucalypt open woodlands
(with a range of dominants including *Eucalyptus phoenicea, E.
kombolgiensis, E. miniata and E. dichromophloia*). Most of the
bioregion is Aboriginal land, including a major part of Kakadu
National Park.

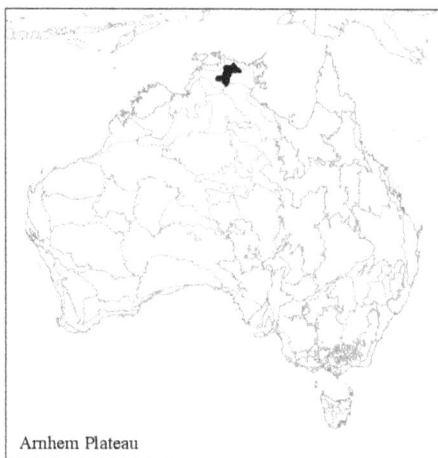

Arnhem Plateau

Species   52   Specimens   149   Adults   149   Larvae   0

	J	F	M	A	M	J	J	A	S	O	N	D	n.d.
Family Lestidae													
Austrolestes insularis											1		
Indolestes alleni											1		
Indolestes obiri						3					1		
Family Isostictidae													
Austrosticta fieldi						1							
Eurysticta coomalie											1		
Lithosticta macra					2	1							
Family Platycnemidae													
Nososticta baroalba					1								
Nososticta coelestina					1								
Nososticta fraterna											2		
Family Coenagrionidae													
Aciagrion fragile						1							
Agriocnemis pygmaea					1	1							

	J	F	M	A	M	J	J	A	S	O	N	D	n.d.
Argiocnemis rubescens						1							
Austroagrion exclamationis						1		4			1		
Austrocnemis maccullochi						1					2		
Ceriagrion aeruginosum								2			4		
Ischnura aurora						1		7					
Ischnura heterosticta					3	2		7			4		
Ischnura pruinescens					3						2		
Family Aeshnidae													
Gynacantha dobsoni					2						2		
Gynacantha nourlangie					2						1		
Hemianax papuensis					1	1							
Family Gomphidae													
Austroepigomphus turneri											5		
Ictinogomphus australis					1						3		
Family Corduliidae													
Pentathemis membranulata											1		
Family Libellulidae													
Agrionoptera i.allogenes										1	1		
Brachydiplax denticauda		1								1	2		
Crocothemis nigrifrons					1						1		
Diplacodes bipunctata					1	1							
Diplacodes haematodes						1							
Diplacodes nebulosa						1					2		
Diplacodes trivialis					1	1		1		1	1		
Hydrobasileus brevistylus					1					1	2		
Lathrecista asiatica festa			1							1	1		
Macrodiplax cora					1						2		
Nannodiplax rubra						1							
Nannophlebia eludens					1								
Nannophlebia mudginberri					2								
Neurothemis stigmatizans					1						1		
Notolibellula bicolor					1								
Orthetrum caledonicum					1								
Orthetrum migratum					1						2		
Orthetrum sabina					1	1					1		
Potamarcha congener										1	1		
Rhodothemis lieftincki											2		
Rhyothemis braganza											2		

	J	F	M	A	M	J	J	A	S	O	N	D	n.d.
Rhyothemis graphiptera					1	1					1		
Rhyothemis phyllis					1								
Tholymis tillarga										1	1		
Tramea loewii										1			
Urothemis aliena				1						1	4		
Zyxomma elgneri										1	1		
Zyxomma petiolatum					1					1	1		

ARC   Arnhem Coast   3,335,669 ha

Coastal strip extending from just east of Cobourg Peninsula to just north of the mouth of the Rose River in southeastern Arnhem Land, and including many offshore islands, most notably Groote Eylandt (and its satellites), the English Company and Wessel group, and the Crocodile Islands. Coastal vegetation includes well developed heathlands, mangroves and saline flats, with some floodplain and wetland areas, most notably the extensive paperbark forest and sedgelands of the Arafura Swamp. Coastal dune systems are unusually well developed on sections of Groote Eylandt and Cape Arnhem Peninsula. Rugged Cretaceous sandstone areas occur on Groote Eylandt and islands of the Wessel group. Tertiary laterites are extensive on the Gove Peninsula. Inland from the coast, the dominant vegetation type is eucalypt tall open forest, typically dominated by Darwin woollybutt (*Eucalyptus miniata*) and Darwin stringybark (*E. tetrodonta*), with smaller areas of monsoon rainforest and eucalypt woodlands.

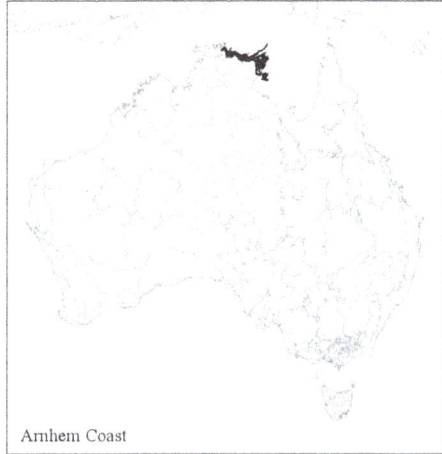

Arnhem Coast

Species   52   Specimens   183   Adults   183   Larvae   0

	J	F	M	A	M	J	J	A	S	O	N	D	n.d.
Family Lestidae													
Indolestes alleni	2												
Indolestes obiri						2							
Lestes concinnus					1						2		
Family Platycnemidae													
Nososticta fraterna											3		
Family Coenagrionidae													
Aciagrion fragile	1					1							
Agriocnemis pygmaea					1	2					2		
Agriocnemis rubricauda								1					
Argiocnemis rubescens						2							

	J	F	M	A	M	J	J	A	S	O	N	D	n.d.
Austroagrion exclamationis	1	2			2								
Ceriagrion aeruginosum	3							1					
Ischnura heterosticta					2	2		1			1		
Pseudagrion jedda					1								
Pseudagrion microcephalum											2		
Family Aeshnidae													
Anax georgius								1					
Anax gibbosulus											1	1	
Anax guttatus	1												
Austrogynacantha heterogena						1							
Gynacantha dobsoni											1		
Gynacantha kirbyi							1						
Gynacantha nourlangie						1							
Gynacantha rosenbergi								1					
Family Gomphidae													
Austroepigomphus turneri	10												
Ictinogomphus australis									1				1
Family Corduliidae													
Hemicordulia kalliste	1												
Pentathemis membranulata	1												
Family Libellulidae													
Agrionoptera i.allogenes	1							1					
Brachydiplax denticauda	1							1			2	1	
Crocothemis nigrifrons								8					
Diplacodes bipunctata	2		2	2									
Diplacodes haematodes					1	1					1		
Diplacodes nebulosa	1												
Diplacodes trivialis	2				1	1		5			1		
Lathrecista asiatica festa	2							2					
Macrodiplax cora								7				1	
Nannodiplax rubra	2		1	3	1	1		1					
Nannophya paulsoni	1												
Neurothemis oligoneura												1	
Neurothemis stigmatizans	3		2					3					
Orthetrum balteatum	1												
Orthetrum caledonicum	1							1			1		
Orthetrum migratum	3												
Orthetrum sabina	1				1			4					

	J	F	M	A	M	J	J	A	S	O	N	D	n.d.
Pantala flavescens	1	2		2									
Potamarcha congener						1				1			
Rhodothemis lieftincki	1				1						1		
Rhyothemis braganza	2												
Rhyothemis graphiptera	1			1		2		6					
Rhyothemis phyllis	1							6		3		1	
Tholymis tillarga						2		1			3		
Tramea loewii	1							2					
Tramea stenoloba	1												
Urothemis aliena	1												

GUC   Gulf Coastal   2,711,718 ha

Gently undulating plains with scattered rugged areas on Proterozoic sandstones and Tertiary sediments; sandy red earths and shallow gravelly, sandy soils; Darwin Stringybark woodland with spinifex understorey.

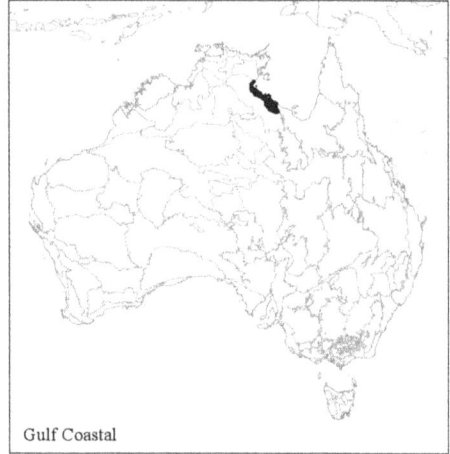

Gulf Coastal

Species   21   Specimens   36   Adults   36   Larvae   0

	J	F	M	A	M	J	J	A	S	O	N	D	n.d.
Family Lestidae													
Austrolestes insularis											2		
Lestes concinnus										2	1	1	
Family Platycnemidae													
Nososticta coelestina			1										
Nososticta fraterna							2				2	1	
Family Coenagrionidae													
Agriocnemis pygmaea											2		
Argiocnemis rubescens							1						
Austroagrion exclamationis											1		
Austroagrion watsoni							1						
Pseudagrion aureofrons							1						
Pseudagrion jedda											1		
Pseudagrion microcephalum											1		
Family Aeshnidae													
Gynacantha dobsoni										1			
Gynacantha nourlangie				1									
Hemianax papuensis		2										1	

	J	F	M	A	M	J	J	A	S	O	N	D	n.d.
Family Libellulidae													
Diplacodes bipunctata	2									1			
Diplacodes haematodes											1		
Nannodiplax rubra											1		
Orthetrum caledonicum		1											
Pantala flavescens	2												
Rhyothemis graphiptera			1						1				
Tramea loewii			1										

GFU   Gulf Fall and Uplands   11,847,909 ha

Undulating terrain with scattered low, steep hills on Proterozoic and Palaeozoic sedimentary rocks, often overlain by lateritised Tertiary material; skeletal soils and shallow sands; Darwin Boxwood and Variable-barked Bloodwood woodland to low open woodland with spinifex understorey.

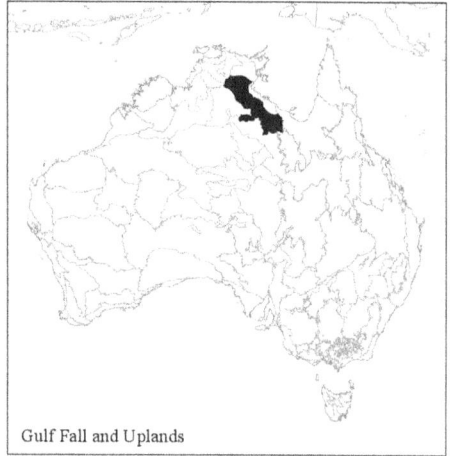

Gulf Fall and Uplands

Species   47   Specimens   257   Adults   255   Larvae   2

	J	F	M	A	M	J	J	A	S	O	N	D	n.d.
Family Lestidae													
Austrolestes insularis											1		
Lestes concinnus				1	2					1	4		
Family Isostictidae													
Austrosticta fieldi				1	1					1			
Austrosticta frater				4									1
Family Platycnemidae													
Nososticta fraterna	1								1	9	4		
Family Coenagrionidae													
Agriocnemis argentea										5			
Agriocnemis pygmaea										3	3		
Argiocnemis rubescens										5	4		
Austroagrion exclamationis										4	4		
Austroagrion watsoni					4					5			
Austrocnemis maccullochi										4	6		
Ceriagrion aeruginosum	1										1		
Ischnura aurora				6	3				1	4			1
Ischnura heterosticta										3	3		
Pseudagrion aureofrons										2			

	J	F	M	A	M	J	J	A	S	O	N	D	n.d.
Pseudagrion cingillum										8	1		
Pseudagrion jedda	1									8	2		
Pseudagrion microcephalum	1				2				3	3	10		
Xanthagrion erythroneurum											1		
Family Aeshnidae													
Gynacantha nourlangie				2						4	1		
Hemianax papuensis	2						1	1					
Family Gomphidae													
Antipodogomphus neophytus	1												
Austroepigomphus turneri											2		2
Austrogomphus mjobergi											1		
Ictinogomphus australis				1						1	2		
Family Corduliidae													
Hemicordulia intermedia	1									7			
Family Libellulidae													
Aethriamanta circumsignata										1			
Crocothemis nigrifrons										1	1		
Diplacodes bipunctata										1	1		
Diplacodes haematodes				8			1		1	3	2		
Macrodiplax cora				1						2	2		
Nannodiplax rubra					3					2	1	1	1
Nannophlebia injibandi										6	1		
Neurothemis oligoneura										1			
Neurothemis stigmatizans										1	1		
Notolibellula bicolor										1	1		
Orthetrum caledonicum	1				1					2			1
Orthetrum migratum					2					4			
Pantala flavescens	1												
Potamarcha congener										1	1		
Rhodothemis lieftincki											2		
Rhyothemis braganza				2						2			
Rhyothemis graphiptera				1						4	3		
Tholymis tillarga											2		
Tramea loewii					1					3			
Tramea stenoloba				1								1	
Zyxomma elgneri										1	2		

219

STU   Sturt Plateau   9,857,531 ha

Gently undulating plains on
lateritised Cretaceous sandstones;
neutral sandy red and yellow
earths; variable-barked Bloodwood
woodland with spinifex understorey.

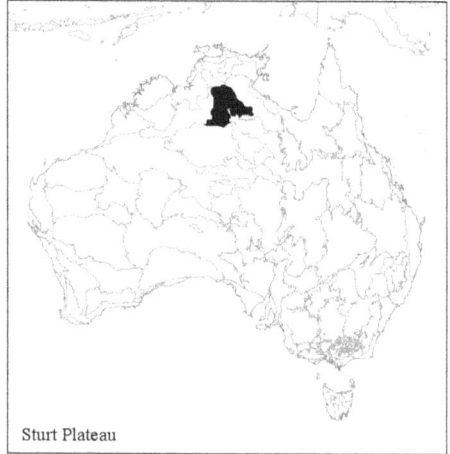

Sturt Plateau

Species   26   Specimens   70   Adults   70   Larvae   0

	J	F	M	A	M	J	J	A	S	O	N	D	n.d.
Family Lestidae													
Lestes concinnus				2			3					1	
Family Isostictidae													
Eurysticta kununurra	1												
Family Platycnemidae													
Nososticta fraterna				1	1							1	
Family Coenagrionidae													
Agriocnemis pygmaea				1									
Austroagrion exclamationis					1								
Ceriagrion aeruginosum						3							
Ischnura aurora				4									
Ischnura heterosticta						1							
Pseudagrion microcephalum					1								
Xanthagrion erythroneurum	1												
Family Aeshnidae													
Anax gibbosulus												1	
Austrogynacantha heterogena				4									
Hemianax papuensis			2	1									

	J	F	M	A	M	J	J	A	S	O	N	D	n.d.
Family Corduliidae													
Pentathemis membranulata												2	
Family Libellulidae													
Crocothemis nigrifrons						1							
Diplacodes bipunctata			2	6									
Diplacodes haematodes			3	1			1						
Diplacodes trivialis						4							
Neurothemis oligoneura						6							
Orthetrum caledonicum			3				1						
Orthetrum sabina						3						1	
Pantala flavescens			1			1							
Rhyothemis graphiptera			1										
Rhyothemis phyllis			1										
Tramea loewii						1							
Tramea stenoloba			1										

BRT   Burt Plain   7,379,719 ha

Plains and low rocky ranges
of Pre-Cambrian granites
with mulga and other acacia
woodlands on red earths.

Burt Plain

Species   12   Specimens   65   Adults   55   Larvae   0

	J	F	M	A	M	J	J	A	S	O	N	D	n.d.
Family Lestidae													
Austrolestes aridus		2			1						1		
Family Coenagrionidae													
Ischnura aurora				1									
Ischnura heterosticta		3											
Xanthagrion erythroneurum	3	16		1					6			2	
Family Aeshnidae													
Hemianax papuensis			1									1	
Family Corduliidae													
Hemicordulia tau				2				1					
Family Libellulidae													
Diplacodes bipunctata			2		2					1			
Diplacodes haematodes				1								1	
Orthetrum caledonicum	3	1	1	1						1		1	
Rhyothemis graphiptera											1		
Tholymis tillarga		1											
Tramea stenoloba	1	2		1								3	

222

MAC   MacDonnell Ranges   3,929,444 ha

High relief ranges and foothills covered with spinifex hummock grassland, sparse acacia shrublands and woodlands along watercourses.

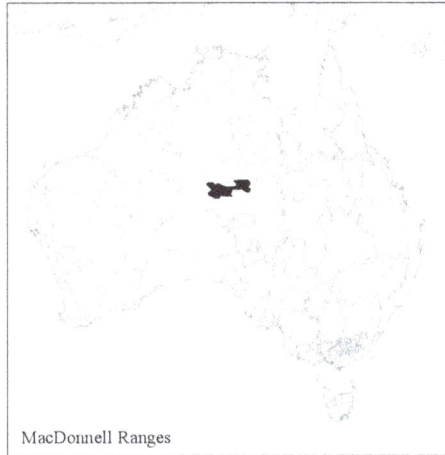

MacDonnell Ranges

Species   20   Specimens   264   Adults   264   Larvae   0

	J	F	M	A	M	J	J	A	S	O	N	D	n.d.
Family Lestidae													
Austrolestes annulosus	3		1										
Austrolestes aridus												1	
Family Coenagrionidae													
Argiocnemis rubescens							1						
Austroagrion watsoni					1				2	1	1		
Ischnura aurora	4	2	4	3		1	1	1	1			2	
Ischnura heterosticta	1	1	3		1		1	2	1	2			
Xanthagrion erythroneurum	6			2	4		1	1		4			1
Family Aeshnidae													
Austrogynacantha heterogena											1		
Hemianax papuensis	3	5	9	6			2	1			1	2	
Family Corduliidae													
Hemicordulia flava	2											3	
Hemicordulia tau			2		2		1						
Family Libellulidae													
Crocothemis nigrifrons			3	2									
Diplacodes bipunctata		2	5	12	1	1	1		5	1			1
Diplacodes haematodes	2	8	23	4	3		1	2	3	12	4		

	J	F	M	A	M	J	J	A	S	O	N	D	n.d.
Orthetrum caledonicum	4	7	15	5			1	2	3	6	4	1	
Orthetrum migratum		1		2							1		
Pantala flavescens		3	2					1	1				
Rhyothemis graphiptera		1											
Tramea loewii		1								2			
Tramea stenoloba			2	1							1		

FIN   Finke   7,267,416 ha

Arid sandplains, dissected
uplands and valleys formed
from Pre-Cambrian volcanics
with spinifex hummock
grasslands and acacia
shrublands on red earths and
shallow sands.

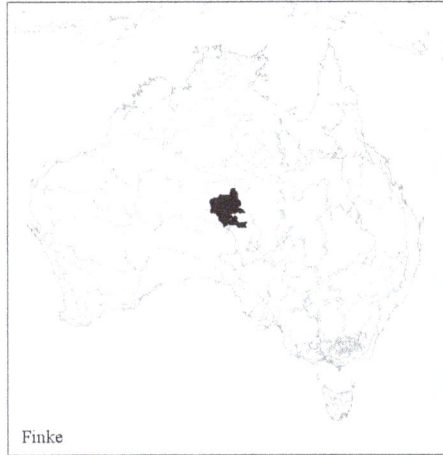

Finke

Species   12   Specimens 53   Adults   53   Larvae   0

	J	F	M	A	M	J	J	A	S	O	N	D	n.d.
Family Lestidae													
Austrolestes aridus								2					
Family Coenagrionidae													
Ischnura aurora									1				
Ischnura heterosticta		2							3	1			
Xanthagrion erythroneurum										1			
Family Aeshnidae													
Hemianax papuensis		1											
Family Corduliidae													
Hemicordulia flava		1											
Hemicordulia tau		2											1
Family Libellulidae													
Diplacodes bipunctata		4											
Diplacodes haematodes		10	1						6				
Orthetrum caledonicum		7	1	3					2		1		
Pantala flavescens		1	1										
Tramea loewii		1											

KIN   King   425,567 ha

Perhumid warm coastal plains
and low hills comprising
King Island and the north-
western tip of Tasmania.
It is a region of subdued
topography and low relief.
Precambrian metamorphic
rocks are overlain by diverse
soils, including recent marine
deposits covered by deep
sandy profiles that support
extensive *Eucalyptus obliqua*
open forest and *Nothofagus
cunninghamii* closed forest.
Acacia melanoxylon closed
forest and *Melaleuca ericifolia*
closed forest occur on poorly drained low-lying sites. The
vegetation of King Island has been substantially degraded by
clearing and burning following European settlement.

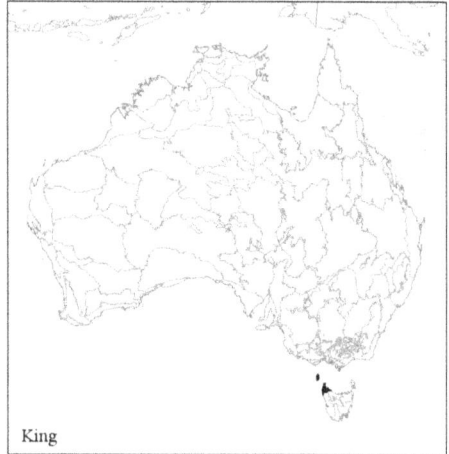

King

Species   12   Specimens   30   Adults   30   Larvae   0

	J	F	M	A	M	J	J	A	S	O	N	D	n.d.
Order Odonata													
Family Lestidae													
Austrolestes analis	1										2		
Austrolestes annulosus													1
Austrolestes cingulatus											1		
Austrolestes io		1											
Family Aeshnidae													
Adversaeschna brevistyla		2										1	
Austroaeschna hardyi		3	1							4			
Austroaeschna parvistigma			1							1			
Austroaeschna unicornis										1			
Family Gomphidae													
Austrogomphus guerini		2											

	J	F	M	A	M	J	J	A	S	O	N	D	n.d.
Family Synthemistidae													
Synthemis tasmanica		1											
Family Corduliidae													
Hemicordulia tau		1									1		
Procordulia jacksoniensis	1	1							3				

FUR   Furneaux   537,543 ha

Moist and dry subhumid warm coastal plains and granitic island chain comprised of the Furneaux islands and coastal north-eastern Tasmania. Devonian granites dominate the elevated areas of the subregion forming low rugged ranges. These are overlain by shallow stony/gravelly gradational or duplex soils carrying *Eucalyptus amygdalina* open forest and woodland with *Eucalyptus nitida* open heath on higher peaks. Quaternary/Tertiary materials overlain by deep sandy soils typify extensive lowland plains, coastal deposits and dunes. Coastal plains have been heavily modified by agriculture (grazing).

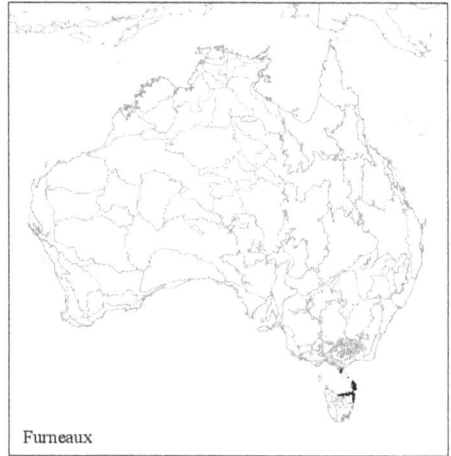

Furneaux

Species   30   Specimens   530   Adults   530   Larvae   0

	J	F	M	A	M	J	J	A	S	O	N	D	n.d.
Family Hemiphlebiidae													
Hemiphlebia mirabilis	11		1								5	4	
Family Synlestidae													
Synlestes weyersii		1											
Family Lestidae													
Austrolestes analis	8	7	3	1	1					5	15	8	1
Austrolestes annulosus	9	3	9	1	2	5	7		6	15	23	9	
Austrolestes cingulatus	1					1							
Austrolestes io	3	1								5	2		
Austrolestes leda									1		1		
Austrolestes psyche	9	6	6				4		2	8	8	11	
Family Argiolestidae													
Austroargiolestes icteromelas		1											
Family Coenagrionidae													
Austroagrion watsoni	2	2	4						4	4	5	2	
Ausrocoenagrion lyelli										1	1	5	
Ischnura aurora	3	5	1						2	5	8	4	

	J	F	M	A	M	J	J	A	S	O	N	D	n.d.
Ischnura heterosticta	6	4	4	1	1	2	2		8	11	13	5	
Xanthagrion erythroneurum	2	1	2							1	5	2	
Family Aeshnidae													
Adversaeschna brevistyla	4	10	8	4	1	5			3	7	10	4	1
Austroaeschna multipunctata	3	3											
Austroaeschna tasmanica				1									
Austroaeschna unicornis										3			
Hemianax papuensis	2	1								2			
Family Gomphidae													
Austrogomphus guerini		1	1						1	1	5	1	
Family Synthemistidae													
Archaeosynthemis orientalis	1												
Eusynthemis guttata	1												
Synthemis eustalacta	3											1	
Synthemis tasmanica			2		1								1
Family Corduliidae													
Hemicordulia tau	3	5	2						3	5	7	3	
Procordulia jacksoniensis	2	3	3			1			1	8	12	3	
Family Libellulidae													
Austrothemis nigrescens	1	1									1		
Diplacodes bipunctata	1	2								1	1	1	
Nannophya dalei	3										6		
Orthetrum caledonicum											1		

TCH   Tasmanian Central Highlands   767,849 ha

Perhumid cool to cold high
plateau surface and rugged
mountain ranges to the west
formed by Jurassic dolerite and
Tertiary basalts, with skeletal
soils to alluvium in valleys,
and humid cool to cold lower
plateau surface underlain
by Jurassic dolerite, Permo-
Triassic sediments and Tertiary
basalts, with sandy to clay
loam soils. Vegetation ranging
from dry sclerophyll woodlands
and wet sclerophyll forest on
the lower plateau to alpine
complexes and coniferous

Tasmanian Central Highlands

forest patches in fertile, fire protected situations on the higher
plateau. Land use is a combination of conservation, forestry,
agriculture (grazing) and water catchment.

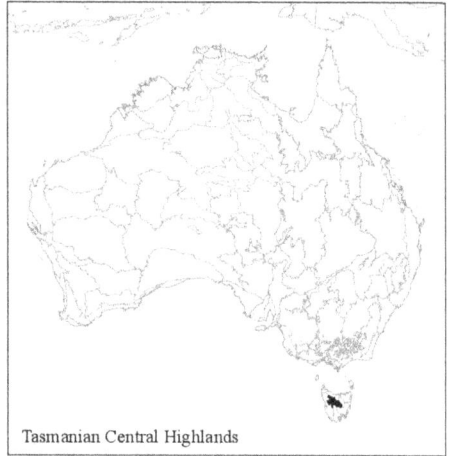

Species   19   Specimens   182   Adults   182   Larvae   0

	J	F	M	A	M	J	J	A	S	O	N	D	n.d.
Family Lestidae													
Austrolestes analis		2											
Austrolestes annulosus	2	2	1								1		
Austrolestes cingulatus	4	7								1	8		
Austrolestes io											1	1	
Austrolestes psyche	7	9									6		
Family Coenagrionidae													
Ausrocoenagrion lyelli											1		
Ischnura aurora	1	2									1		
Ischnura heterosticta	1												1
Family Austropetaliidae													
Archipetalia auriculata	4										2	2	
Family Aeshnidae													
Austroaeschna hardyi	15	11	2	1						1			
Austroaeschna parvistigma	2	9	1		1						3		
Austroaeschna tasmanica	1	3	1							2			

	J	F	M	A	M	J	J	A	S	O	N	D	n.d.
Austroaeschna unicornis													1
Hemianax papuensis													1
Family Gomphidae													
Austrogomphus guerini	1												1
Family Synthemistidae													
Syn. gomphomacromioides	7	4											1
Synthemis tasmanica	9	22	1								2	2	
Family Corduliidae													
Hemicordulia tau		4	1								2		
Procordulia jacksoniensis	1	1											1

BEL   Ben Lomond   657,500 ha

Humid cool/cold mountain ranges situated in Tasmania's inland north-east. The mountains are capped by Jurassic dolerite with shallow gradational soils. Silurian-Devonian siltstones and mudstones covered with gradational soils constitute a substantial part of the lower hills. Lowland vegetation comprising mainly open sclerophyll woodlands and heath while the upper slopes consist of wet sclerophyll forests, some rainforest and alpine vegetation in the highest regions. Land use: forestry, mining and agriculture (grazing).

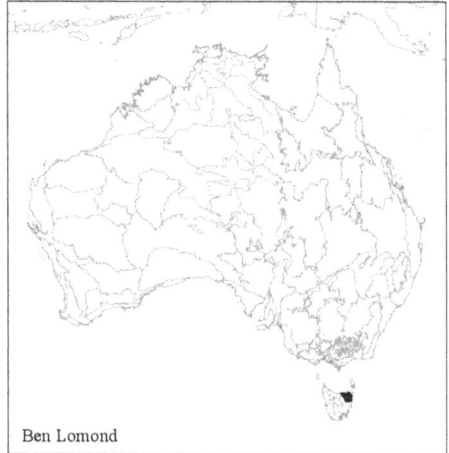

Ben Lomond

species 13   Specimens 34   Adults   34   Larvae   0

	J	F	M	A	M	J	J	A	S	O	N	D	n.d.
Family Hemiphlebiidae													
Hemiphlebia mirabilis		1											
Family Lestidae													
Austrolestes analis	2	2									3		
Austrolestes annulosus	2	1									1		
Austrolestes io												1	
Austrolestes psyche		1								1	1		
Family Aeshnidae													
Austroaeschna tasmanica										1			
Austroaeschna unicornis		1											
Family Gomphidae													
Austrogomphus guerini		1								1		1	
Family Synthemistidae													
Syn. gomphomacromioides		1								1			

	J	F	M	A	M	J	J	A	S	O	N	D	n.d.
Synthemis tasmanica	2	3											
Family Corduliidae													
Hemicordulia tau		1											
Procordulia jacksoniensis		1										2		
Family Libellulidae													
Nannophya dalei												2	

TNM Tasmanian Northern Midlands 415,445 ha

Dry subhumid cool inland lowland plain underlain by Tertiary basalts, Jurassic dolerite, Permo-Triassic sand - stones, and recent alluvium lying in the Tamar. Vegetation comprises grasslands and grassy woodlands on deep loams and alluvium and dry sclerophyll forest and woodland on Tertiary. Grasslands and woodlands have been reduced to remnants. Land use is primarily agriculture (grazing) with some forestry. Extensively cleared for agriculture.

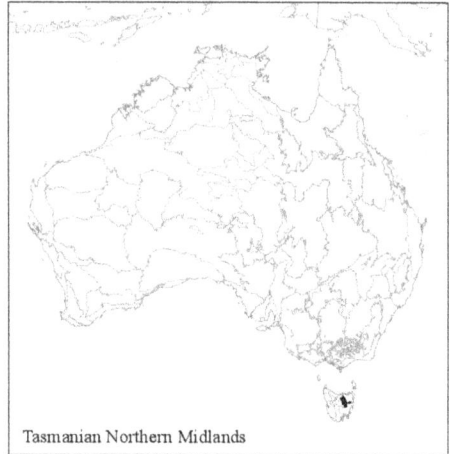

Tasmanian Northern Midlands

Species   16   Specimens   61   Adults   61   Larvae   0

	J	F	M	A	M	J	J	A	S	O	N	D	n.d.
Family Lestidae													
Austrolestes analis	2	1											
Austrolestes annulosus	2										1		
Austrolestes psyche	2												
Family Coenagrionidae													
Austroagrion watsoni	3												
Ausrocoenagrion lyelli	6												
Ischnura aurora	1									1	1	1	
Ischnura heterosticta	1												
Family Aeshnidae													
Adversaeschna brevistyla	1	1			1						1	1	
Austroaeschna parvistigma	1											1	
Austroaeschna unicornis									4	2			
Family Gomphidae													
Austrogomphus guerini	1									4	1		
Family Synthemistidae													
Synthemis tasmanica	3		3										

234

	J	F	M	A	M	J	J	A	S	O	N	D	n.d.
Family Corduliidae													
Hemicordulia tau	2									1		1	
Procordulia jacksoniensis	6				1								
Family Libellulidae													
Austrothemis nigrescens	2												
Nannophya dalei	1												

TSE   **Tasmanian South East**   1,131,822 ha

Subhumid cool to subhumid
warm coastal plains on a highly
indented coastline, bordered
inland by low mountain ranges
formed from Jurassic dolerite
and Permo-Triassic sediments.
Soils predominantly clay to
sandy loams. Vegetation is
predominantly dry sclerophyll
forest, with patches of wet
sclerophyll forest, relict
rainforest, coastal heath
and dry coniferous forest.
Extensive areas have been
converted to improved pasture
and cropland. Land is use
primarily agriculture (grazing) and forestry.

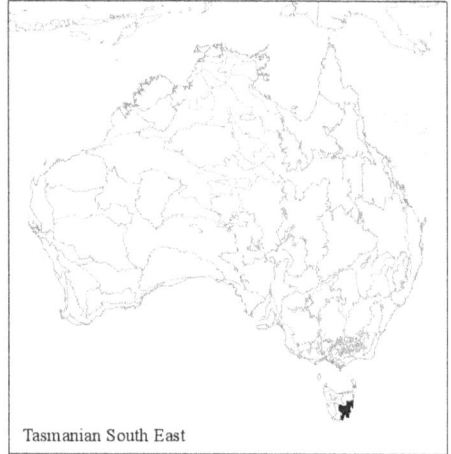

Tasmanian South East

Species   23   Specimens   129   Adults   129   Larvae   0

	J	F	M	A	M	J	J	A	S	O	N	D	n.d.
Family Lestidae													
Austrolestes analis	1	2										1	
Austrolestes annulosus	11	7									1	1	
Austrolestes cingulatus		1											
Austrolestes io		1							2				
Austrolestes psyche		8							1			1	
Family Coenagrionidae													
Austroagrion watsoni		1											
Ausrocoenagrion lyelli										1			
Ischnura aurora	2	2									1	1	
Ischnura heterosticta	3	2									1	1	
Xanthagrion erythroneurum		2									1		
Family Aeshnidae													
Adversaeschna brevistyla	4	3	2	1								2	1
Austroaeschna hardyi			4							1			
Austroaeschna parvistigma	2	3	7	1								1	1
Austroaeschna tasmanica	1												

	J	F	M	A	M	J	J	A	S	O	N	D	n.d.
Austroaeschna unicornis		1											
Hemianax papuensis		1											
Family Gomphidae													
Austrogomphus guerini	2	6										1	1
Family Synthemistidae													
Archaeosynthemis orientalis	4												1
Family Corduliidae													
Hemicordulia australiae											1		
Hemicordulia tau	2	6	2										
Procordulia jacksoniensis	2	4										1	1
Family Libellulidae													
Austrothemis nigrescens		1											
Nannophya dalei	1											1	

TSR   Tasmanian Southern Ranges   757,228 ha

Humid cool mountainous tract of central southern Tasmania. Permo-Triassic sediments and Jurassic dolerite, mantled with sandy to clay loams. Heavily forested, grading from mixed forest, wet sclerophyll forest and patches of rainforest in the uplands to dry sclerophyll forest on the coastal lowlands. Land use primarily forestry and agriculture (grazing and cropping).

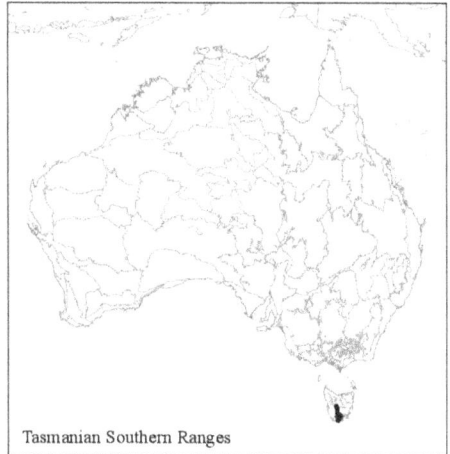

Tasmanian Southern Ranges

Species   19   Specimens   120   Adults   120   Larvae   0

	J	F	M	A	M	J	J	A	S	O	N	D	n.d.
Family Lestidae													
Austrolestes analis	2	1											
Austrolestes annulosus			1										
Austrolestes cingulatus		15										3	
Austrolestes io			3										
Austrolestes psyche		12	1									1	
Family Coenagrionidae													
Ausrocoenagrion lyelli	1												
Ischnura aurora		1											
Ischnura heterosticta													1
Family Austropetaliidae													
Archipetalia auriculata												3	1
Family Aeshnidae													
Adversaeschna brevistyla	1												
Austroaeschna hardyi	5	12	1	1									
Austroaeschna parvistigma		7	1										
Austroaeschna tasmanica	1	6	1										
Austroaeschna unicornis			1										

	J	F	M	A	M	J	J	A	S	O	N	D	n.d.
Family Gomphidae													
Austrogomphus guerini	1	2											
Family Synthemistidae													
Syn. gomphomacromioides	2	3											
Synthemis eustalacta													
Synthemis tasmanica	3	16	1										1
Family Corduliidae													
Hemicordulia tau		1									2	1	
Procordulia jacksoniensis			1									1	

TWE   Tasmanian West   1,565,077 ha

Perhumid cold lowlands, low hills and low ranges, comprising most of coastal and inland western Tasmania. Folding and subsequent erosion has resulted in rugged dissected inland ranges dominated by Precambrian and Cambrian rocks supporting oligotrophic acid peat soils or shallow organic horizons over deep mineral profiles. From 300 metres elevation a discontinuous coastal plain slopes westward to the ocean. Vegetation is a complex mosaic of rainforest (*Nothofagus*), buttongrass (*Gymnoschoenus sphaerocephalus*) moorlands and *Eucalyptus nitida* scrub. Principal land uses are conservation, mining and forestry.

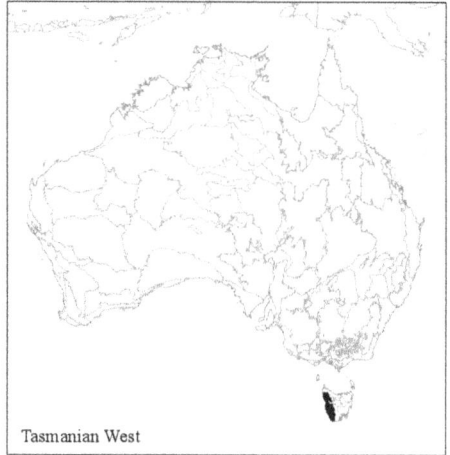

Tasmanian West

Species   21   Specimens   198   Adults   198   Larvae   0

	J	F	M	A	M	J	J	A	S	O	N	D	n.d.
Family Lestidae													
Austrolestes annulosus	1									1			
Austrolestes cingulatus	1												
Austrolestes psyche	8	7	2							3		1	
Family Coenagrionidae													
Ischnura aurora		1											
Ischnura heterosticta	2												
Xanthagrion erythroneurum										1			
Family Austropetaliidae													
Archipetalia auriculata		1											
Family Aeshnidae													
Adversaeschna brevistyla	1	3											
Austroaeschna hardyi	10	16	7	1	2					1	1	1	
Austroaeschna parvistigma	1	7	6								1		
Austroaeschna tasmanica		1								1			
Austroaeschna unicornis		2		1							1	1	

	J	F	M	A	M	J	J	A	S	O	N	D	n.d.
Hemianax papuensis										1			
Family Gomphidae													
Austrogomphus guerini	2	5	1									1	
Family Synthemistidae													
Archaeosynthemis orientalis											1		
Syn. gomphomacromioides	3	25	6										
Synthemis tasmanica	10	23	3									2	
Family Corduliidae													
Hemicordulia australiae		1											
Hemicordulia tau	1	4	3										
Procordulia jacksoniensis	2	7								1		1	
Family Libellulidae													
Nannophya dalei	1									1			

Köppen Climate Zones for Australia

Desert - Hot (persistently dry)	315
Desert - Hot (Summer drought)	246
Desert - Hot (Winter drought)	246
Desert Warm - (persistently dry)	246
Equatorial - Rainforest (monsoonal)	246
Equatorial - Savanna	283
Grassland - Hot (persistently dry)	302
Grassland - Hot (Summer drought)	307
Grassland - Hot (Winter drought)	295
Grassland - Warm (persistently dry)	311
Grassland - Warm (Summer drought)	246
Subtropical - Distinctly dry summer	313
Subtropical - Distinctly dry winter	279
Subtropical - Moderately dry winter	275
Subtropical - No dry season	248

Temperate - Distinctly dry (and hot) Summer	309
Temperate - Distinctly dry (and mild) Summer	246
Temperate - Distinctly dry (and warm) Summer	299
Temperate - Moderately dry Winter (hot Summer)	305
Temperate - Moderately dry Winter (warm Summer)	246
Temperate - No dry season (cool Summer)	317
Temperate - No dry season (mild Summer)	291
Temperate - No dry season (warm Summer]	265
Temperate - No dry season (hot Summer)	270
Tropical - Rainforest (monsoonal)	260
Tropical - Rainforest (persistently wet)	287
Tropical - Savanna	254

The seven climate zones with the fewest number of species, ranging from 16 to one, do not warrant individual checklists and flight time charts. From the maps below it can be seen that they comprise desert areas or regions that are small or are predominantly offshore. A table is given showing the species recorded for each of these zones.

For those zones with greater numbers of records, collection dates of specimens of larvae cannot contribute to flight calendars for adults so they have been ignored. However, on the rare occasion where a larva species has been sampled without a comparable adult being recorded, that larva is included in the total number of species.

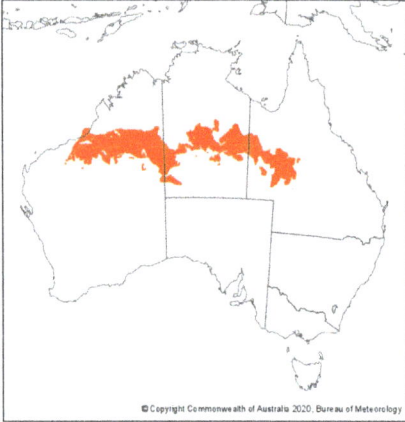

Desert - hot (winter drought)

Equatorial - rainforest (monsoonal)

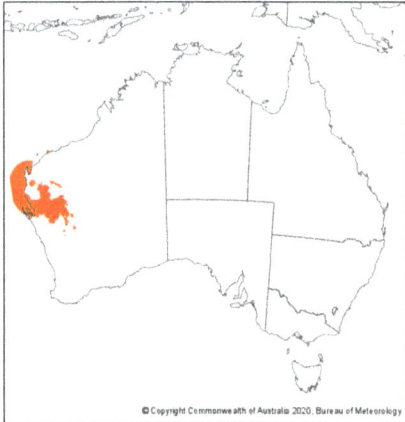

Desert - hot (summer drought)

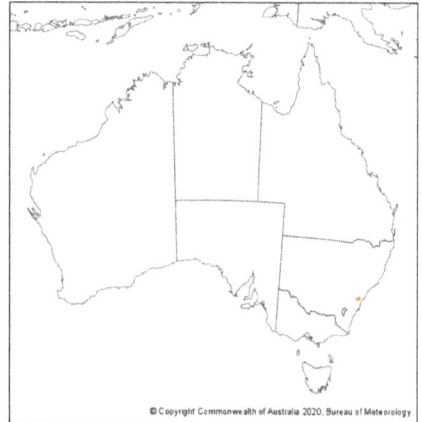

Temperate - moderately dry winter (warm summer)

Grassland - warm (summer drought)

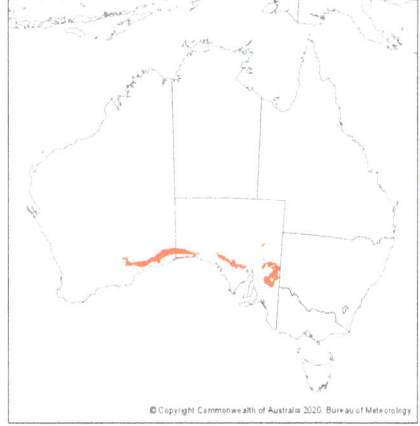

Desert - warm (persistently dry)

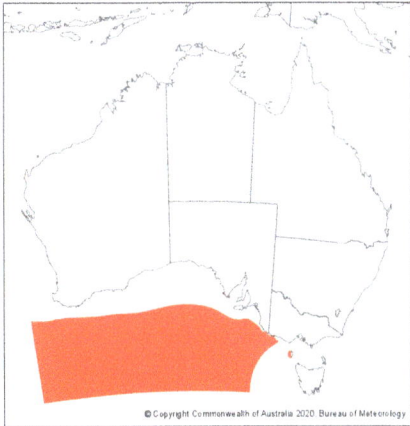

Temperate - distinctly dry (and mild) summer

	A	B	C	D	E	F	G
Family Synlestidae							
Synlestes weyersii				1			
Family Lestidae							
Austrolestes aridus	4		4		1		
Austrolestes io			1				
Family Argiolestidae							
Austroargiolestes icteromelas				1			
Family Isostictidae							
Oristicta filicicola		2					
Rhadinosticta simplex				1			
Family Coenagrionidae							
Argiocnemis rubescens		8					
Ischnura aurora	2				3		
Ischnura heterosticta	1	1		1			
Pseudagrion aureofrons	1						
Pseudagrion ignifer		3					
Pseudagrion microcephalum	1						
Teinobasis rufithorax		1					
Xanthagrion erythroneurum	5		12		2		
Family Aeshnidae							
Adversaeschna brevistyla				1		1	1
Austroaeschna obscura				1			
Austroaeschna unicornis					1		
Dendroaeschna conspersa				1			
Hemianax papuensis	7		16		4	2	
Family Gomphidae							
Austroepigomphus gordoni			2				
Austrogomphus prasinus		3					
Family Synthemistidae							
Choristhemis flavoterminata				3			
Parasynthemis regina				3			
Family Corduliidae							
Hemicordulia australiae				1			
Hemicordulia tau	5		6		2	3	
Family Libellulidae							
Diplacodes bipunctata	12		26		2	3	
Diplacodes haematodes	13	1	3		2		
Macrodiplax cora	1						

	A	B	C	D	E	F	G
Nannodiplax rubra		1					
Nannophlebia eludens		1					
Orthetrum caledonicum	9		7		4	1	
Orthetrum serapia		4					
Orthetrum villosovittatum		1					
Pantala flavescens	3		4				
Tholymis tillarga	1						
Tramea loewii	1		7				
Tramea stenoloba	1		1				
Zyxomma petiolatum		1					
Species	16	12	12	11	9	5	1
Specimens	67	27	89	15	21	10	1

A   Desert - Hot (Winter drought)
B   Equatorial - Rainforest (monsoonal)
C   Desert - Hot (Summer drought)
D   Temperate - Moderately dry Winter (warm Summer)
E   Desert - Warm (persistently dry)
F   Grassland - Warm (Summer drought)
G   Temperate - Distinctly dry (and mild) Summer

Subtropical - No dry season

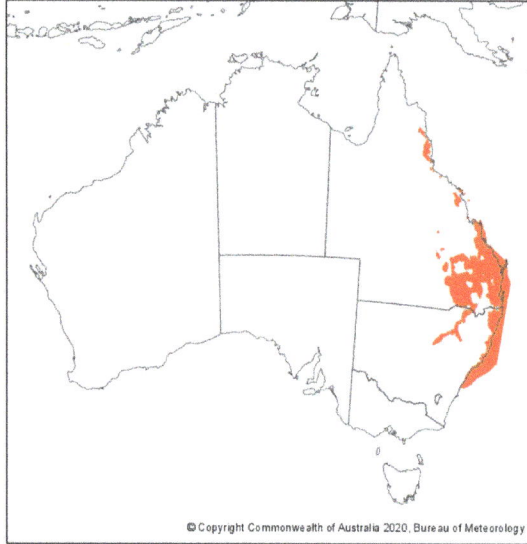

Subtropical - no dry season

Species   189   Specimens   10251   Adults   9979   Larvae   272

	J	F	M	A	M	J	J	A	S	O	N	D	n.d.
Family Synlestidae													
Chorismagrion risi		1	4	3	4		2			2	7	1	2
Episynlestes albicauda	6	8	13	24	15	6	1	2	1	9	1	21	5
Episynlestes cristatus	10	3	5	9							1		1
Episynlestes intermedius	1									1	4	1	1
Synlestes selysi	10	2	11	5	18	2				3	1	11	2
Synlestes tropicus	1	5	8	1	9		1			3	5	11	1
Synlestes weyersii		2		3					1				1
Family Lestidae													
Austrolestes analis	1	2											
Austrolestes aridus				2		1				1	1	3	2
Austrolestes cingulatus								1			2		
Austrolestes io											1		
Austrolestes leda	27	14	29	34	37	4	7	10	35	20	35	51	6
Austrolestes minjerriba	58	3	17	17	14			2	39	10	7	40	2

	J	F	M	A	M	J	J	A	S	O	N	D	n.d.
Austrolestes psyche	12	8		7	2	1			1	16	3	5	1
Indolestes tenuissimus												1	
Lestes concinnus			1	2						2			
Family Argiolestidae													
Austroargiolestes amabilis	8	6								8	13	13	
Austroargiolestes aureus	1	2	4	1				1		6	9	25	4
Austroargiolestes chrysoides	4	2	12			1			28	12	4	15	
Austroargiolestes elke	1									5	6		1
Austroargiolestes icteromelas	61	32	56	55	28	1	6	3	25	64	72	73	2
Griseargiolestes albescens	24	7	4	5	8	1		2	34	22	4	20	2
Griseargiolestes eboracus	3												
Griseargiolestes fontanus	2	1								1	1	6	
Griseargiolestes metallicus			1								1	4	
Podopteryx selysi	2		2	3							1	1	
Family Lestoideidae													
Diphlebia coerulescens	29	5	29	3	4					13	20	27	
Diphlebia euphoeoides	17	2	9	28				2	4	3	15	11	
Diphlebia hybridoides		1								1	10	22	
Diphlebia lestoides	6								1	3	11	9	1
Diphlebia nymphoides	2								6	1	4	3	
Lestoidea barbarae											6	2	
Lestoidea brevicauda											2		
Lestoidea conjuncta	3	2	4	3						2	11	6	1
Lestoidea lewisiana											2	8	
Family Isostictidae													
Labidiosticta vallisi	5	3	3	11	10					4	2	1	
Neosticta canescens	10									5	13	3	
Neosticta fraseri				3						1	6	8	
Neosticta silvarum												2	
Oristicta filicicola		2	1									1	
Rhadinosticta banksi	2		3	1									
Rhadinosticta simplex	18	7	4	2	8				3	1	2	10	
Family Platycnemidae													
Nososticta coelestina			8							1			
Nososticta solida	40	10	9	17	1	1		1	5	16	32	49	
Nososticta solitaria	11	2	25	6	1				3	3	12	16	2
Family Coenagrionidae													
Aciagrion fragile	7			1			1		3	3	7		

	J	F	M	A	M	J	J	A	S	O	N	D	n.d.
Agriocnemis argentea	2	1	1	4	1					2	8	8	
Agriocnemis pygmaea	29	20	6	31	1				17	12	18	18	1
Agriocnemis rubricauda	1								1	1	2		
Archibasis mimetes												1	
Argiocnemis rubescens	18	10	17	18	2	1	1		20	13	14	26	2
Austroagrion exclamationis	12		3	1		1				1	1	12	
Austroagrion watsoni	13	7	10	5	7	2			43	44	45	21	2
Austrocnemis splendida	8	10	2	2	7				5	19	9	13	
Ausrocoenagrion lyelli													1
Caliagrion billinghursti										1			
Ceriagrion aeruginosum	21	6	12	6	1	1				6	12	14	
Ischnura aurora	22	10	13	7	6	4		1	30	14	34	34	1
Ischnura heterosticta	54	24	30	30	26	8		3	31	31	59	39	4
Ischnura pruinescens	10		1	1		1			7	1			
Pseudagrion aureofrons	11	8	3	8	2	2			1	7	7	23	
Pseudagrion cingillum	1			1		2			1		2	2	
Pseudagrion ignifer	22	12	32	43	9	1		1	3	13	20	40	
Pseudagrion microcephalum	14	17	17	19	6	5		3	7	11	21	37	4
Teinobasis rufithorax			1										
Xanthagrion erythroneurum	2	4	4	4	1	1		1	1	3	21	7	2
Family Aeshnidae													
Acanthaeschna victoria										7	2		
Adversaeschna brevistyla	23	13	7	6	3	2		3	8	4	14	16	3
Anax gibbosulus		4	1	1					1				
Anax guttatus	1												
Antipodophlebia asthenes	5										2	4	
Austroaeschna christine	3		1	1						2			
Austroaeschna cooloola	3	1							1	2	4		2
Austroaeschna eungella	2			1					1	1	3		
Austroaeschna pinheyi	4	2	5		4	1			1				
Austroaeschna pulchra	6	1	7	2	2				1		4		1
Austroaeschna sigma	8	3	5	4	2				3		1	4	
Austroaeschna subapicalis		1											
Austroaeschna unicornis	2			1									
Austrogynacantha heterogena	13	10	6	13	1				1		4	8	1
Austrophlebia costalis	6	1						1		1	3	9	2
Austrophlebia subcostalis	1							1			2	1	
Dendroaeschna conspersa			3		2								

	J	F	M	A	M	J	J	A	S	O	N	D	n.d.
Dromaeschna forcipata	1	6	4								5	3	
Dromaeschna weiskei	8		7	5	2						3	4	
Gynacantha dobsoni		1							1			2	
Gynacantha mocsaryi												1	
Gynacantha rosenbergi		1								1			3
Hemianax papuensis	10	12	22	21	9	3	3	2	5	7	3	17	5
Notoaeschna geminata									1	1			1
Spinaeschna watsoni									2				
Telephlebia cyclops	9	4	4	1						1	3	11	
Telephlebia godeffroyi													
Telephlebia tillyardi		3								1			
Telephlebia tryoni	4	4									6	9	6
Family Petaluridae													
Petalura gigantea	7										2	6	2
Petalura ingentissima	4	1										1	1
Petalura litorea	16	1	1							2	1	10	2
Family Gomphidae													
Antipodogomphus acolythus	1		4	1								3	
Antipodogomphus proselythus	3	1									1	4	1
Austroepigomphus praeruptus	5	1	3	3				1	1	7	1	8	1
Austroepigomphus turneri		1								3	3	1	
Austrogomphus amphiclitus	47	15	14	27	8				3	11	16	34	4
Austrogomphus australis	1									1			
Austrogomphus bifurcatus	1	8	1	2									
Austrogomphus cornutus	7		4	2						1	6	5	1
Austrogomphus divaricatus	2	3		2						1	2		
Austrogomphus doddi		1											
Austrogomphus guerini	4												
Austrogomphus ochraceus	14	2						1	12	13	21	40	1
Austrogomphus prasinus	3		5	1						1	1		
Hemigomphus atratus										1			
Hemigomphus comitatus										1	3	1	
Hemigomphus cooloola	5	1									1	6	
Hemigomphus gouldii	4							1		4	6	3	
Hemigomphus heteroclytus	12	2	4	3					3	7	15	15	1
Hemigomphus theischingeri	1		1							1	1		
Ictinogomphus australis	33	13	10	7	2						2	16	3
Odontogomphus donnellyi										2			

	J	F	M	A	M	J	J	A	S	O	N	D	n.d.
Zephyrogomphus longipositor												2	
Family Synthemistidae													
Archaeosynthemis orientalis	1												
Choristhemis flavoterminata	44	12	15	7	4		1			7	32	56	1
Eusynthemis aurolineata			2	1						1	3	4	
Eusynthemis barbarae												5	
Eusynthemis cooloola												1	
Eusynthemis netta												3	1
Eusynthemis nigra	55	4	9	5				2		5	36	46	18
Eusynthemis rentziana	1								1			1	1
Eusynthemis tenera										1		1	
Parasynthemis regina	2	5		1			1			3	6	5	1
Synthemis eustalacta	1												
Tonyosynthemis claviculata		3	1										
Family Macromiidae													
Macromia tillyardi		1											
Family Corduliidae													
Hemicordulia australiae	59	22	18	30	12	2		1	23	29	46	43	4
Hemicordulia continentalis	25	15	7	2	2				2	3	8	21	
Hemicordulia intermedia			5	2	1				1		2		
Hemicordulia superba	2	1	1	3							2		3
Hemicordulia tau	4	2		5	1	1			2	7	5	15	3
Procordulia jacksoniensis											1	1	
Family Libellulidae													
Aethriamanta circumsignata	1	1							2		1	1	
Aethriamanta nymphaea	3	3	5		1					1	1	3	
Agrionoptera i.allogenes	4	13	4	2	7					1		1	
Agrionoptera l.biserialis		1	3						1		1	1	
Austrothemis nigrescens	9	6			1				7	12	2	16	2
Brachydiplax denticauda	15	15	13	1	2					2	6	10	
Crocothemis nigrifrons	16	20	16	26	7	3	1		10	7	14	9	4
Diplacodes bipunctata	49	45	80	83	40	10	3	16	46	20	39	38	8
Diplacodes haematodes	44	15	44	58	27	5	2	10	14	26	25	40	2
Diplacodes melanopsis	9	16	8	2	1			1	1	26	12	11	7
Diplacodes trivialis	2	8	9	4	5				4	1	1	1	
Hydrobasileus brevistylus	11	8	6	1			2				2	10	5
Lathrecista asiatica festa		1	6		1							1	
Macrodiplax cora	4	5	3	3	2	1			3			2	

	J	F	M	A	M	J	J	A	S	O	N	D	n.d.
Nannodiplax rubra	36	6	21	6	1			2	29	1	9	33	1
Nannophlebia eludens	3	2	18	5	1					1	3	1	
Nannophlebia risi	22	7	12	17	3				1	9	5	16	2
Nannophya australis	48	4	5	4	6			1	35	22	5	49	
Nannophya dalei	1												
Neurothemis stigmatizans		2	4	8	3		2	1	5	3	7	3	
Orthetrum boumiera	46	10	31	6	1					7	5	36	1
Orthetrum caledonicum	50	42	32	54	15	2		2	14	16	35	45	3
Orthetrum migratum	1		1								1	1	
Orthetrum sabina	29	21	27	17	2	1		1	4	7	10	12	
Orthetrum serapia		1						1					
Orthetrum villosovittatum	58	50	49	53	20	2	1	2	17	17	40	53	4
Pantala flavescens	5	6	16	7	1			1	3		1	1	4
Potamarcha congener					1								
Rhodothemis lieftincki	15	12	9	4				1		3	1	9	
Rhyothemis braganza	3						1				2		
Rhyothemis graphiptera	45	16	18	8	4			2	2	7	22	27	5
Rhyothemis phyllis	41	21	5		1				1	2	14	34	3
Rhyothemis princeps	1						1	1					
Rhyothemis respledens												3	
Tholymis tillarga		1		1						1		1	
Tramea eurybia	44	3	1	2				1			5	32	1
Tramea loewii	43	22	33	17	7			1	20	22	14	17	5
Tramea stenoloba	6			1									
Urothemis aliena	3			1						3	2	1	
Zyxomma elgneri	27	10	6	1						1	1	5	
Genera incertae sedis													
Archaeophya magnifica											1	5	
Austrocordulia refracta	8	2								2	3	3	3
Austrophya mystica		1								1	1	1	
Cordulephya bidens			2		1								
Cordulephya pygmaea	1		3	8	11								6
Lathrocordulia garrisoni											2		
Micromidia atrifrons	4	8	1	1							1	1	1
Micromidia convergens	6	2									2		
Pseudocordulia circularis											1		
Pseudocordulia elliptica	5									1	4	2	

Tropical – Savanna

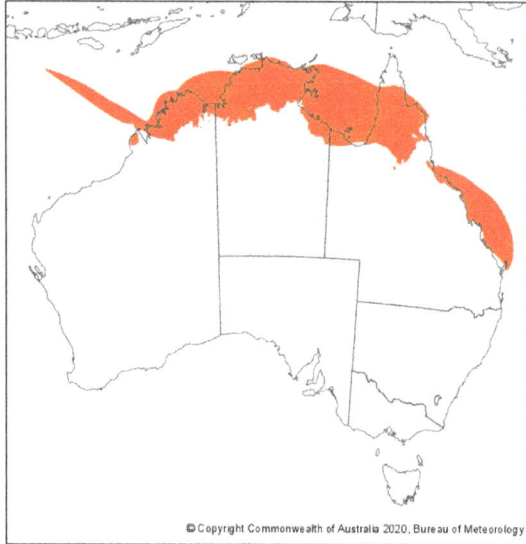

Tropical - savanna

Species   155   Specimens   4828   Adults   4828   Larvae   0

	J	F	M	A	M	J	J	A	S	O	N	D	n.d.
Family Synlestidae													
Chorismagrion risi									1				
Episynlestes albicauda	1												
Synlestes selysi			1		3								
Synlestes tropicus					1						1		
Family Lestidae													
Austrolestes insularis	10	2	4	4	2	2	8			2	6	2	
Austrolestes leda	3												
Indolestes alleni	3						1			1	2	2	
Indolestes obiri			1		6	5					4		
Indolestes tenuissimus	2		1				1				1	6	
Lestes concinnus	4	8	5	7	13	1	7	1	1	7	14	8	
Family Argiolestidae													
Austroargiolestes aureus											6	1	

	J	F	M	A	M	J	J	A	S	O	N	D	n.d.
Austroargiolestes icteromelas	1									1			
Family Lestoideidae													
Diphlebia euphoeoides	3	3	3	3	4			1	1	1	8	13	1
Lestoidea conjuncta										2			
Lestoidea lewisiana												1	
Family Isostictidae													
Austrosticta fieldi		1	2		3	2					3		
Austrosticta frater		4	14	1									
Austrosticta soror		1	1										
Eurysticta coomalie	1	4	2	1							2		
Eurysticta reevesi			4										
Labidiosticta vallisi										1			
Lithosticta macra			3	1	3	1							
Neosticta fraseri											1		
Oristicta filicicola		1	2		1					2	4	4	
Oristicta rosendaleorum			6										
Rhadinosticta banksi	1	13	23	1	1								
Rhadinosticta simplex					1								
Family Platycnemidae													
Nososticta baroalba	1	7	11	2	11		9		1	5	13	7	
Nososticta coelestina	1		8	7	10		2	1	3	3	5	9	
Nososticta fraterna	4	2	2	1	22	1	13	7	4	2	34	12	2
Nososticta kalumburu		1				1	3	21					
Nososticta koolpinyah	1				2						1		10
Nososticta koongarra			4		8	1		1			17		
Nososticta liveringa							2	72					
Nososticta mouldsi					7						1		
Nososticta solida	3	1	8	1					4	8	7	1	
Nososticta solitaria		2	12	6	1		3	1	2	8	31	15	
Nososticta taracumbi										1			
Family Coenagrionidae													
Aciagrion fragile	9	1	2	8	4	2	1	2	1		17	2	3
Agriocnemis argentea			2				1	1			1		
Agriocnemis dobsoni			1										
Agriocnemis pygmaea		1	3	9	9	8	7	8	3	2	17	1	
Agriocnemis rubricauda	1		5		1		7	1	4	2	2	2	
Archibasis mimetes	1				1			1			1		
Argiocnemis rubescens			6	2	14	5	7	9	2	2	8	5	

	J	F	M	A	M	J	J	A	S	O	N	D	n.d.
Austroagrion exclamationis	2	6	8		14	7	12	14	2	6	37	10	
Austroagrion watsoni	3						2	8			1	2	
Austrocnemis maccullochi		2			14	9	1	9		11	11	9	
Austrocnemis obscura							1	1					1
Austrocnemis splendida							2						
Ceriagrion aeruginosum	9		1	11	4		9	5	3	6	17	11	1
Ischnura aurora	1	2	7	5	16	8	15	19	8	20	32	6	1
Ischnura heterosticta	1	2	9	11	25	20	5	23	3	12	56	16	1
Ischnura pruinescens	2		3	11	11	2	9	10	4	2	10	5	7
Pseudagrion aureofrons							1		1				
Pseudagrion cingillum		9			5	1					8	5	
Pseudagrion ignifer	1		7	2	7	1	2	3	2	4	1	14	
Pseudagrion jedda	1	11			51		19	1		1	12	4	1
Pseudagrion lucifer			3		7	1	10	7	1		1		
Pseudagrion microcephalum	1	3	16		38	10	16	8	1	5	24	17	
Xanthagrion erythroneurum	1		1								1		
Family Aeshnidae													
Anaciaeschna jaspidea				1									
Anax georgius								2					
Anax gibbosulus	1	2	1	2	2				1	3	4	1	
Anax guttatus	9			1	1			2			1	5	
Austrogynacantha heterogena	2	8	2	5	7	1					1	1	1
Dromaeschna forcipata	3		1								5	4	
Gynacantha dobsoni			2		3		2	4	1	4	6	1	
Gynacantha kirbyi			1			3				2			
Gynacantha mocsaryi	1					1					1	1	1
Gynacantha nourlangie			2		18	15	14	16	2	4	6		1
Gynacantha rosenbergi								1		4			
Hemianax papuensis	4	10	11	2	6	3	5	2		1	3	4	4
Spinaeschna watsoni											1		
Family Petaluridae													
Petalura ingentissima	3	1											
Petalura pulcherrima												1	
Family Gomphidae													
Antipodogomphus acolythus		1											
Antipodogomphus dentosus		4											
Antipodogomphus edentulus												3	
Antipodogomphus neophytus		4									1	1	

	J	F	M	A	M	J	J	A	S	O	N	D	n.d.
Antipodogomphus proselythus	1										2	3	2
Austroepigomphus praeruptus	1												
Austroepigomphus turneri	10	1		1					4	2	2	4	
Austrogomphus amphiclitus			3	2									
Austrogomphus arbustorum											1	2	2
Austrogomphus divaricatus											2	1	
Austrogomphus mjobergi	4	19		1							1		
Austrogomphus prasinus	3	1	2	11							1	9	7
Hemigomphus comitatus	6	1									3	14	
Hemigomphus magela				2							9		
Hemigomphus theischingeri			1										
Ictinogomphus australis	1	2		3	4		1	1	2		15	7	5
Family Synthemistidae													
Choristhemis flavoterminata	7	3	4	1							5	16	
Eusynthemis nigra	1										15		
Tonyosynthemis claviculata			1								2	5	
Family Macromiidae													
Macromia tillyardi	1										2	2	
Macromia viridescens											1		
Family Corduliidae													
Hemicordulia australiae	2												
Hemicordulia continentalis			1										
Hemicordulia intermedia	2	4	8	2	1	1	2	1	1		5	10	
Hemicordulia kalliste	1												
Hemicordulia tau					1						1	1	
Pentathemis membranulata	1	2									11	8	
Family Libellulidae													
Aethriamanta circumsignata		1	3	1	1						1	1	
Aethriamanta nymphaea						1		3				8	
Agrionoptera i.allogenes	2		2		5		3	1	2	1	3	3	
Agrionoptera l.biserialis			2						1			2	
Brachydiplax denticauda	2	2	3	6	5		9	3	1	3	19	7	1
Camacinia othello	1	1									1		1
Crocothemis nigrifrons	1		1	2	6	5		9	2	4	7	11	
Diplacodes bipunctata	25	11	28	18	27	8	5	10	1	4	2	4	1
Diplacodes haematodes	9	7	27	21	26	13	19	12	13	17	15	12	
Diplacodes nebulosa	4	3	5	3	3	2	7	5	1	1	7	4	
Diplacodes trivialis	13		9	8	14	10	9	10	4	17	12	12	1

	J	F	M	A	M	J	J	A	S	O	N	D	n.d.
Hydrobasileus brevistylus	3		5	3	1				4	3	9	3	
Lathrecista asiatica festa	6	1	7	3	3	5		2	1	8	7	7	1
Macrodiplax cora		1	7	1	7		1	8	2	3	10	6	
Nannodiplax rubra	9	6	9	6	16	19	31	15	16	7	22	13	1
Nannophlebia eludens	3		6	12	10	1	2	1		1	9	3	
Nannophlebia injibandi		1		8	5		1	1		1			1
Nannophlebia mudginberri			4		7	1	2	1		2	3		
Nannophlebia risi			1			1		1			3		
Nannophya australis		1			1						1		
Nannophya paulsoni	1												
Neurothemis oligoneura												1	
Neurothemis stigmatizans	14	9	19	17	26	12	22	17	15	12	22	23	3
Notolibellula bicolor		1		2	1	2							
Orthetrum balteatum	1										1	4	
Orthetrum boumiera												8	
Orthetrum caledonicum	4	8	6	7	10	2	8	2	1	13	6	6	1
Orthetrum migratum	6	6	8	1	5	3	5	6	4	2	7	7	
Orthetrum sabina	4	1	9	3	7	6	5	14	4	4	10	2	2
Orthetrum serapia			2	1	1	1	1			4	1		
Orthetrum villosovittatum	7	1	8		6		1		2	6	5	4	
Pantala flavescens	9	6	12	7	8	7	5	3	1	3	2	3	4
Potamarcha congener	2	3	5		2	1	6		2	2	4	1	
Rhodothemis lieftincki	6		3	4	3		5		3	2	11	9	
Rhyothemis braganza	5	1	5	1	1			2	2	2	10	13	1
Rhyothemis graphiptera	6	10	18	9	9	7	8	16	9	11	7	12	
Rhyothemis phyllis	3		1	3	1			7		7	5	6	
Rhyothemis princeps	8	1	3	6	1						3	15	1
Rhyothemis resplendens	1		1								3	2	
Tetrathemis i.cladophila	1		2								1	1	
Tholymis tillarga	2	7	7	3	9	2	2	3	1	1	13	7	
Tramea loewii	5	5	12	9	10	6	8	9	2	7	13	13	
Tramea stenoloba	1			2	1	1				1		2	
Urothemis aliena	1				3			1		1	4	4	
Zyxomma elgneri	1	4		1	8			1	1	2	5	3	
Zyxomma petiolatum					1					1	1		
Genera incertae sedis													
Austrocordulia refracta											1	2	
Austrocordulia territoria					9								

	J	F	M	A	M	J	J	A	S	O	N	D	n.d.
Austrophya mystica		1										1	
Cordulephya bidens				1									
Micromidia·atrifrons	2	9				2							

Tropical - Rainforest (monsoonal)

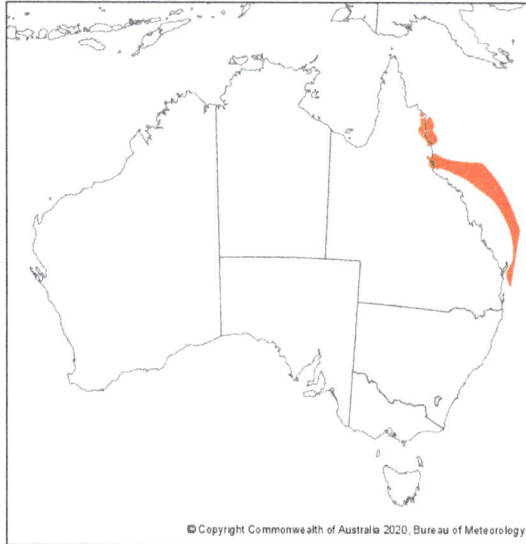

Tropical - rainforest (monsoonal)

Species   127   Specimens   2341   Adults   2341   Larvae   0

	J	F	M	A	M	J	J	A	S	O	N	D	n.d.
Family Synlestidae													
Chorismagrion risi	2			6						5	3		
Synlestes selysi			8	1									
Synlestes tropicus	3			5						3	2		1
Family Lestidae													
Austrolestes insularis										2			
Indolestes alleni	5			1			1	1		3			
Indolestes tenuissimus	5			2					5	2	1	2	
Lestes concinnus	1			2		1		3			1		
Family Argiolestidae													
Austroargiolestes aureus	10	2	3						3	7	7	5	
Austroargiolestes icteromelas			6										
Podopteryx selysi	2	1	1	1						2	1	2	
Family Lestoideidae													

	J	F	M	A	M	J	J	A	S	O	N	D	n.d.
Diphlebia euphoeoides	12	1	45	17	4			3	26	14	12	10	1
Diphlebia hybridoides	5	1									2	7	
Lestoidea brevicauda	1		1							1			
Lestoidea conjuncta	14	3	5	6	1			3	8	1	5		1
Lestoidea lewisiana												3	
Family Isostictidae													
Neosticta fraseri	1		1		1					1			
Oristicta filicicola	8		5	11				1	1	3	2	1	
Rhadinosticta banksi			2										
Rhadinosticta simplex	1									2			
Family Platycnemidae													
Nososticta coelestina	4	2	5	10	2				1	8	1	2	
Nososticta solida	3		1		1			6			2	3	
Nososticta solitaria	10	2	13	10			1	3	6	12	4	16	2
Family Coenagrionidae													
Aciagrion fragile	2			4			1	4	3		2		
Agriocnemis argentea	3			2		2			5	2	3		
Agriocnemis dobsoni										1	1		
Agriocnemis pygmaea	1		1	2	1					3	1		
Agriocnemis rubricauda										3	1		
Archibasis mimetes	2			1						1			1
Argiocnemis rubescens	1		3	5	1		1	3	4	10	2	4	6
Austroagrion exclamationis											2		
Austroagrion watsoni								2			2		
Ceriagrion aeruginosum	4		6	1	3		2			14	7	5	4
Ischnura aurora	1		11		1			3	4				
Ischnura heterosticta	2		3						10		1		
Ischnura pruinescens	9		2		1		1		8	2			1
Pseudagrion cingillum	1								1				
Pseudagrion ignifer	9	1	7	6	6			2	22	11	13		6
Pseudagrion jedda	1												
Pseudagrion lucifer		1											
Pseudagrion microcephalum	3	1	5	10	1			2	11	8	8	9	2
Teinobasis rufithorax				1						1			
Family Aeshnidae													
Adversaeschna brevistyla	1							1				1	
Anax gibbosulus	1							1	1				
Anax guttatus	9												1

	J	F	M	A	M	J	J	A	S	O	N	D	n.d.
Austroaeschna speciosa													1
Austrogynacantha heterogena	4	4		1									
Dromaeschna forcipata	14		2	3						2	5	5	2
Dromaeschna weiskei	2		1	6						1	1	1	
Gynacantha dobsoni	5		2		1			7	1	5	3	1	2
Gynacantha kirbyi			1										
Gynacantha mocsaryi	8		2						2		1	2	
Gynacantha nourlangie				2									
Gynacantha rosenbergi	5	1	2		2				1	8	6	6	2
Hemianax papuensis	3	1		2				2					1
Telephlebia tillyardi	1		1							3	1		
Family Petaluridae													
Petalura ingentissima	9												
Petalura pulcherrima		4											1
Family Gomphidae													
Antipodogomphus acolythus	1												
Antipodogomphus neophytus	1												
Antipodogomphus proselythus	6	3									3		
Austroepigomphus praeruptus				1					1				
Austroepigomphus turneri		2									2		
Austrogomphus amphiclitus	1												
Austrogomphus arbustorum	1										3		3
Austrogomphus bifurcatus	2			2							2		
Austrogomphus cornutus	1												
Austrogomphus divaricatus		1		1						1	2		
Austrogomphus doddi		2											1
Austrogomphus prasinus	24	4	8	10						3	11	3	
Hemigomphus comitatus	2	3								1	1	2	
Hemigomphus heteroclytus			1								1		
Hemigomphus theischingeri			1								3	1	
Ictinogomphus australis	6	2									1	4	2
Family Synthemistidae													
Choristhemis flavoterminata	12		4				1			1	3	8	
Choristhemis olivei											1		
Eusynthemis nigra	4	2	1	1						3	5	8	
Tonyosynthemis claviculata	1												
Family Macromiidae													
Macromia tillyardi	7					1					2		

	J	F	M	A	M	J	J	A	S	O	N	D	n.d.
Family Corduliidae													
Hemicordulia australiae			1					2		1	1	2	
Hemicordulia continentalis	4	2	1							1			
Hemicordulia intermedia	5	2						1			4	1	
Hemicordulia tau	1											1	
Family Libellulidae													
Aethriamanta circumsignata												1	
Aethriamanta nymphaea			2						1				
Agrionoptera i.allogenes	2	1	4	7	3					4		5	
Agrionoptera l.biserialis	14		7	2	2		1	1	6	3	3		
Brachydiplax denticauda	3	1	3	2						5	5	2	1
Brachydiplax duivenbodei	1		1	1						1		2	
Camacinia othello	4												
Crocothemis nigrifrons	4				1					5	2		1
Diplacodes bipunctata	8		10	5		2	1	1	3		2	2	
Diplacodes haematodes	10		14	11	1	1	1	1	15	4	4	4	1
Diplacodes nebulosa			1		1								
Diplacodes trivialis	7	1	16	6	6	3	4	3	4	1	2		
Hydrobasileus brevistylus	3								1	3	1	2	
Lathrecista asiatica festa	13		3	4	1			2	1	2	1	5	1
Macrodiplax cora	1							1	1	2	1		
Nannodiplax rubra	7			9	4	1	1	3	7	4	6	5	2
Nannophlebia eludens	8	1	10	9	6	2				4	4	7	2
Nannophlebia risi	1		5	2					4		2	1	
Neurothemis stigmatizans	16	2	17	17	14	22	4	22	9	13	2	17	
Notolibellula bicolor		3											1
Orthetrum caledonicum	2		2	10	2			1	3		2		
Orthetrum migratum	1												
Orthetrum sabina	3		5	6	3	4	2	9	11	5	1	1	
Orthetrum serapia	6		3	3				3				1	
Orthetrum villosovittatum	8	4	8	4	4	2	1	4	8	11	6	6	1
Pantala flavescens	9		6	17	1	5		6				1	1
Rhodothemis lieftincki	1			2	1	1			2	5	1	2	
Rhyothemis braganza	4								1		1	6	
Rhyothemis graphiptera	3		2			1			2	2	1	1	
Rhyothemis phyllis	19		3	2	2				2	4	7	3	1
Rhyothemis princeps	4		14						1	3		2	1
Rhyothemis resplendens	16	1									4	6	1

	J	F	M	A	M	J	J	A	S	O	N	D	n.d.
Tetrathemis i.cladophila	4		3	1	4			1		5	1	2	
Tholymis tillarga	5	1	9	2		2		2	1	2	1	1	2
Tramea loewii	4	1	2	7	1	1		7	5			4	1
Tramea propinqua	1												
Zyxomma elgneri			1									1	
Zyxomma petiolatum	5		3	1								2	
Genera incertae sedis													
Archaeophya magnifica	2										2	2	
Austrocordulia refracta		1								2	8	4	
Austrophya mystica	2										1	1	
Cordulephya bidens				2	2								
Micromidia·atrifrons	2											2	
Pseudocordulia circularis	1											1	
Pseudocordulia elliptica	1												

Temperate - No dry season (warm Summer)

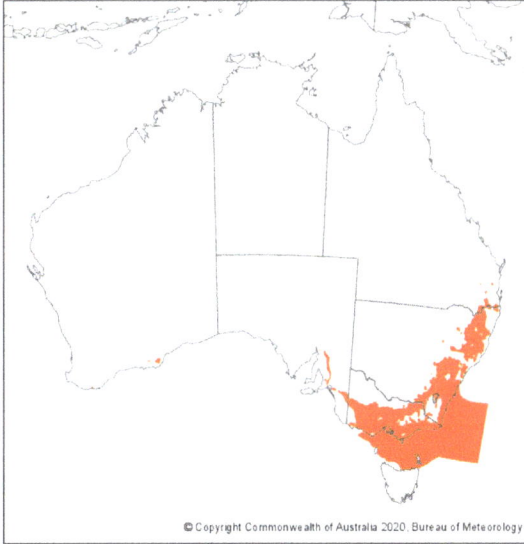

Temperate - no dry season (warm summer)

Species   125   Specimens   16,214   Adults   14,429   Larvae   1,985

	J	F	M	A	M	J	J	A	S	O	N	D	n.d.
Family Hemiphlebiidae													
Hemiphlebia mirabilis	20	1									7	48	41
Family Synlestidae													
Episynlestes albicauda	12		5	6	3	1	1	1			2	7	1
Synlestes selysi	18	7	12	22	13	6			1		4	8	1
Synlestes weyersii	177	90	78	49	11	1	1		2	4	58	101	10
Family Lestidae													
Austrolestes analis	94	42	24	6	3					5	81	133	6
Austrolestes annulosus	28	19	12	8	1	1			8	35	22	45	2
Austrolestes aridus	2		1	1				1	2	8		20	
Austrolestes cingulatus	67	36	15	8						20	41	39	3
Austrolestes io	1	1	1	1			1		1	8	4	10	
Austrolestes leda	64	17	25	8	4	3	1	11	48	68	80	134	10

	J	F	M	A	M	J	J	A	S	O	N	D	n.d.
Austrolestes minjerriba										1	1	1	
Austrolestes psyche	78	15	4	11	3				4	23	47	35	
Family Argiolestidae													
Austroargiolestes alpinus	1	3											
Austroargiolestes amabilis	11		1	1		1				1	5	24	2
Austroargiolestes brookhousei	10						1				2	5	1
Austroargiolestes calcaris	15	3								8	32	42	
Austroargiolestes christine	2	1					1				1	14	1
Austroargiolestes chrysoides	3	1	2							1	5	1	
Austroargiolestes icteromelas	332	110	57	25	2	1		2	18	55	226	385	28
Austroargiolestes isabellae	3							3	5	14	17	6	
Griseargiolestes albescens			1						1			5	
Griseargiolestes bucki	5									2	22		10
Griseargiolestes eboracus	63	4	1	1			1			4	7	18	4
Griseargiolestes fontanus	11			1							2	14	
Griseargiolestes griseus	28	17	2	2					1	4	24	17	10
Griseargiolestes intermedius	9										3	10	4
Family Lestoideidae													
Diphlebia coerulescens	5	2	2	1			3			2	4	10	
Diphlebia lestoides	29	2					1		5	12	61	101	10
Diphlebia nymphoides	75	17	2	1						6	76	34	6
Family Isostictidae													
Neosticta canescens	9	1							1	1	20	38	18
Rhadinosticta simplex	28	30	22	12						2	3	10	
Family Platycnemidae													
Nososticta solida	65	17	6	2		1				2	52	35	1
Family Coenagrionidae													
Agriocnemis pygmaea	4	2	2		1						1	4	
Argiocnemis rubescens	4	2	2	1	1						2	1	1
Austroagrion watsoni	113	51	27	11					14	9	62	110	7
Austrocnemis splendida	5	4								1	4	7	4
Ausrocoenagrion lyelli	1	2									6	28	
Caliagrion billinghursti	3	5									16	13	4
Ischnura aurora	65	23	24	6	3				7	55	90	108	3
Ischnura heterosticta	124	81	39	14	7		1	1	7	25	122	81	6
Pseudagrion aureofrons	6	5	13	4							30	8	1
Pseudagrion ignifer	10	7	1	3							17	1	5
Pseudagrion microcephalum	4		1	1							5	1	1

	J	F	M	A	M	J	J	A	S	O	N	D	n.d.
Xanthagrion erythroneurum	48	31	21	8	2			2	49	43	63	5	
Family Austropetaliidae													
Austropetalia patricia					1			1	15	5		2	
Austropetalia tonyana			1					2	3	17	3	2	
Family Aeshnidae													
Acanthaeschna victoria												4	
Adversaeschna brevistyla	91	34	20	9				2	23	66	102	10	
Antipodophlebia asthenes	1									1	8	1	
Austroaeschn atrata	8	4	1	1						1	8	2	
Austroaeschna flavomaculata		2											
Austroaeschna inermis		2	8	2						1	1		1
Austroaeschna ingrid	20		3								5	24	
Austroaeschna multipunctata	13	28	28	14	4						3	8	1
Austroaeschna obscura	13	8	4	5	4					1	10	12	3
Austroaeschna parvistigma	12	11	5	5							3	11	
Austroaeschna pinheyi	2	2	2										
Austroaeschna pulchra	36	24	54	35	3	1				1	3	21	11
Austroaeschna sigma	18	2	9	8	7	2				1	3	11	23
Austroaeschna subapicalis	51	22	5	3		1	1			1	2	6	
Austroaeschna unicornis	30	33	39	23					1	3		10	7
Austrogynacantha heterogena	1				1		1						
Austrophlebia costalis	10	7	3				1		1	4	4	11	4
Dendroaeschna conspersa			15	12	2				1		1	4	4
Hemianax papuensis	47	34	31	7				6	24	26	30	38	2
Notoaeschna geminata	5	1		1					1	6	9	4	
Notoaeschna sagittata	12	9	1							3	13	28	2
Spinaeschna tripunctata	6	3								4	11	11	
Telephlebia brevicauda	33	12	7	1		2					1	12	2
Telephlebia cyclops	12	8	1		1				1			2	1
Telephlebia godeffroyi	24	12	6	2							1	9	24	10
Family Petaluridae													
Petalura gigantea	99	17									39	90	7
Family Gomphidae													
Austroepigomphus praeruptus	12	2		2								2	
Austrogomphus amphiclitus	34	10	4	2						1	7	14	1
Austrogomphus australis	3										5	5	
Austrogomphus cornutus	8	4	10	1						1	2	16	1
Austrogomphus guerini	162	85	22	4						5	51	104	9

	J	F	M	A	M	J	J	A	S	O	N	D	n.d.
Austrogomphus ochraceus	96	70	18	8					3	34	89		12
Hemigomphus gouldii	79	28	8	1					2	15	69		13
Hemigomphus heteroclytus	30	13	9	5	1			2	2	35	19		3
Ictinogomphus australis													1
Family Synthemistidae													
Archaeosynthemis orientalis	48	3	4					2	9	13			2
Choristhemis flavoterminata	38	14	2	3	1				6	30			1
Eusynthemis aurolineata	29	5	2	2		1			6	26			3
Eusynthemis brevistyla	69	19				1			22	77			1
Eusynthemis guttata	50	29	29	7					14	42			1
Eusynthemis nigra	10	1		1					9	5			
Eusynthemis rentziana	1								1	6			2
Eusynthemis tillyardi	8	8	1	2				4	13	21			4
Eusynthemis ursa									1	1			
Eusynthemis ursula									4	5			
Eusynthemis virgula	24	16	3		3			1	1	22	36		10
Parasynthemis regina	9	4	2	1					2	5			
Synthemis eustalacta	86	35	16	4			1		2	14	57		9
Tonyosynthemis ofarrelli	5									2			
Family Corduliidae													
Hemicordulia australiae	91	48	27	10	3			2	2	30	78		2
Hemicordulia intermedia	1	2	1						1	4			
Hemicordulia superba									2	2			
Hemicordulia tau	71	40	45	19	5	1		21	62	84	128		5
Procordulia jacksoniensis	18	4	2					6	5	14			
Family Libellulidae													
Austrothemis nigrescens	1									1			
Crocothemis nigrifrons	1	3	1						1	2			1
Diplacodes bipunctata	109	56	50	27	7		2	66	115	113	88		8
Diplacodes haematodes	46	35	21	21	3		1	8	67	27			2
Diplacodes melanopsis	49	17	18	4				11	58				6
Hydrobasileus brevistylus	1												
Nannodiplax rubra	1												
Nannophlebia risi	11	5	4	1	1			3	4	11			1
Nannophya australis	33	4	2			1		2	5	3			
Nannophya dalei	43	4	1					5	30	42			1
Orthetrum caledonicum	141	68	39	14			1	7	42	100			1
Orthetrum sabina	1	1	2	1									

	J	F	M	A	M	J	J	A	S	O	N	D	n.d.
Orthetrum villosovittatum	40	15	4	50	1						5	17	2
Pantala flavescens	2		3	2							1		
Rhyothemis graphiptera	4											1	1
Tramea loewii	7	3	3	3					1	4	1	1	
Genera incertae sedis													
Apocordulia macrops	14												2
Archaeophya adamsi		1										3	2
Austrocordulia leonardi										1	8	4	1
Austrocordulia refracta	5								2	5	14	9	
Cordulephya divergens			1	1	1								
Cordulephya montana	4	3	2								1	5	3
Cordulephya pygmaea	5	11	36	30	7	5					1	3	8
Micromidia·atrifrons												1	1
Micromidia convergens				1									

Temperate - No dry season (hot Summer)

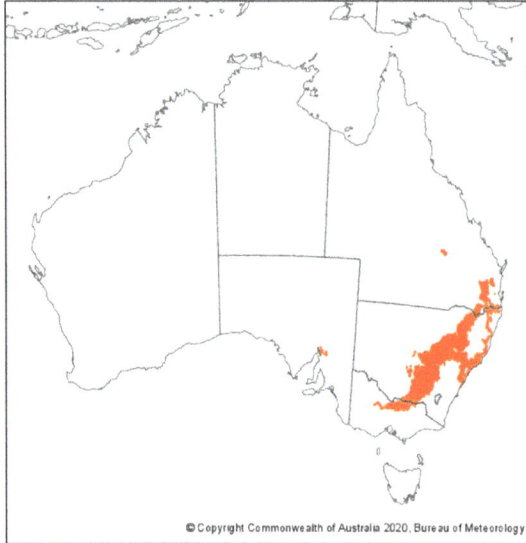

Temperate - no dry season (hot summer)

Species   123   Specimens   4482   Adults   3223   Larvae   1259

	J	F	M	A	M	J	J	A	S	O	N	D	n.d.
Family Synlestidae													
Episynlestes albicauda	6	4	2	3	5					2	1	5	
Synlestes selysi	2		1	4	1	1						5	1
Synlestes weyersii	9	12	5	5					1	1	3	7	9
Family Lestidae													
Austrolestes analis	7	4	3	3						2	5	10	
Austrolestes annulosus	2	2	1										
Austrolestes aridus				2					2		1		
Austrolestes cingulatus	2											2	
Austrolestes leda	30	8	8	10				7	20	14	17	7	3
Austrolestes psyche	13	2	3	1	1					2	4	2	
Lestes concinnus												1	

	J	F	M	A	M	J	J	A	S	O	N	D	n.d.
Family Argiolestidae													
Austroargiolestes amabilis	3	1								1	1	1	
Austroargiolestes brookhousei	1											1	
Austroargiolestes calcaris											2		
Austroargiolestes christine												1	
Austroargiolestes chrysoides	4	1		1							6	3	
Austroargiolestes icteromelas	44	17	11	17	2				14	34	30	23	9
Austroargiolestes isabellae								1			2	1	
Griseargiolestes eboracus		1											1
Griseargiolestes fontanus	2	1											
Griseargiolestes griseus	1		2								1	1	16
Family Lestoideidae													
Diphlebia coerulescens	9	2	2	1							5	2	
Diphlebia lestoides	1			1						1	1	1	4
Diphlebia nymphoides	12	11	2	2					2	18	4	20	
Family Isostictidae													
Labidiosticta vallisi	1				1								1
Neosticta canescens	5	1									13	4	2
Rhadinosticta simplex	7	6	8	1						2	4	2	
Family Platycnemidae													
Nososticta solida	41	24	13	4	1					15	24	16	1
Nososticta solitaria	1												
Family Coenagrionidae													
Agriocnemis pygmaea	4	7		3	1					2	9	15	
Agriocnemis rubricauda												1	
Argiocnemis rubescens	3	1		1						4	5	1	1
Austroagrion watsoni	5	9	4	4	2				7	7	8	9	3
Austrocnemis splendida	3		2								1	1	
Caliagrion billinghursti				1									
Ceriagrion aeruginosum	1	1										1	1
Ischnura aurora	19	7	7	10				1	2	12	3	12	3
Ischnura heterosticta	26	21	17	14	4			1	8	26	11	9	16
Ischnura pruinescens		1											
Pseudagrion aureofrons	12	12	9	8	1					3	10	25	3
Pseudagrion ignifer	7	7	1	7					1		5	1	1
Pseudagrion microcephalum				1							3	5	
Xanthagrion erythroneurum	8	3	15	2	1			1	1	4	13	4	

	J	F	M	A	M	J	J	A	S	O	N	D	n.d.
Family Austropetaliidae													
Austropetalia patricia													1
Family Aeshnidae													
Acanthaeschna victoria												3	
Adversaeschna brevistyla	7	5	7	3	1				2	2	6	6	4
Antipodophlebia asthenes		1										1	
Austroaeschna muelleri		2	1	1					1	3	1	4	1
Austroaeschna multipunctata											1		
Austroaeschna obscura	1												
Austroaeschna parvistigma						1					1		
Austroaeschna pinheyi	1	2	3	2	5						1	3	
Austroaeschna pulchra	5	2		12	2	1			1	1		3	1
Austroaeschna sigma	3	3		5	4	1						7	2
Austroaeschna subapicalis											1		
Austroaeschna unicornis	2	7	1	7									1
Austrogynacantha heterogena	2	1	1										2
Austrophlebia costalis	1									1	3	1	1
Dendroaeschna conspersa			1	7	2								
Hemianax papuensis	8	9	21	3	3		1	3	4	18	8	3	2
Notoaeschna geminata	1											1	
Spinaeschna tripunctata										1		1	1
Telephlebia cyclops	2	4		6		1					1	2	
Telephlebia godeffroyi	3						1			1	2	2	1
Telephlebia undia												1	
Family Petaluridae													
Petalura gigantea	22	1								2	3	11	4
Petalura litorea													1
Family Gomphidae													
Antipodogomphus acolythus	2											6	
Austroepigomphus praeruptus	11	1	2							1	6	7	4
Austrogomphus amphiclitus	19	7	5	2						1	1	24	3
Austrogomphus angelorum	1												
Austrogomphus australis	8	2	3								3	9	3
Austrogomphus cornutus	9	5	20	2						6	12	28	2
Austrogomphus guerini	4	8								4		1	
Austrogomphus ochraceus	9	8	2						1		7	11	6
Hemigomphus gouldii	6	4								1		2	
Hemigomphus heteroclytus	22	11	9	9	2				1	7	15	25	4

	J	F	M	A	M	J	J	A	S	O	N	D	n.d.
Ictinogomphus australis	4	2										2	1
Family Synthemistidae													
Archaeosynthemis orientalis												1	
Choristhemis flavoterminata	19	5	2	4							10	11	3
Eusynthemis aurolineata	7	2		2						2	2	8	2
Eusynthemis brevistyla		2										1	
Eusynthemis deniseae		1								4		7	2
Eusynthemis guttata												1	
Eusynthemis nigra	8	6		1						2	12	34	7
Eusynthemis rentziana											1		
Eusynthemis tillyardi											1	1	1
Eusynthemis virgula	2									1	1	2	3
Parasynthemis regina	15	6	2								1	6	2
Synthemis eustalacta	1	4										1	
Tonyosynthemis ofarrelli	2											1	2
Family Corduliidae													
Hemicordulia australiae	18	8	7	7	1				1	3	10	4	6
Hemicordulia continentalis	1	1										1	
Hemicordulia intermedia	1									1	1	8	
Hemicordulia superba	4	1	1									4	
Hemicordulia tau	9	5	10	3	11				3	13	18	2	2
Procordulia jacksoniensis		1											
Family Libellulidae													
Aethriamanta circumsignata												1	
Aethriamanta **nymphaeae**		1											
Brachydiplax denticauda			1									2	
Crocothemis nigrifrons	4		5	1		1				4		7	
Diplacodes bipunctata	11	16	23	17	5	1		1	2	16	16	16	4
Diplacodes haematodes	31	12	27	25	4				12	25	24	4	1
Diplacodes melanopsis	4	1		2					1	1	4	10	1
Hydrobasileus brevistylus	3	1										1	
Nannodiplax rubra										1	2	1	1
Nannophlebia risi	17	5	3	3						7	4	3	2
Nannophya australis	1										3	4	2
Nannophya dalei										2			
Orthetrum caledonicum	20	27	13	2			1		3	18	15	9	4
Orthetrum sabina	1	3	1						1	4			
Orthetrum villosovittatum	16	4	6	10	1			1		4	4	5	41

	J	F	M	A	M	J	J	A	S	O	N	D	n.d.
Pantala flavescens	1		2										
Rhodothemis lieftincki		1										1	1
Rhyothemis graphiptera	8	1	1								1	10	1
Rhyothemis phyllis	1			1								2	
Tramea loewii		1	6	1				1	3	5	4	2	
Zyxomma elgneri				1									1
Genera incertae sedis													
Apocordulia macrops	1											1	1
Austrocordulia leonardi												1	
Austrocordulia refracta	4		1							1		2	
Cordulephya pygmaea		1	9	5	3	1							1
Micromidia·atrifrons			2	1									
Micromidia convergens	1	1										1	

Subtropical - Moderately dry winter

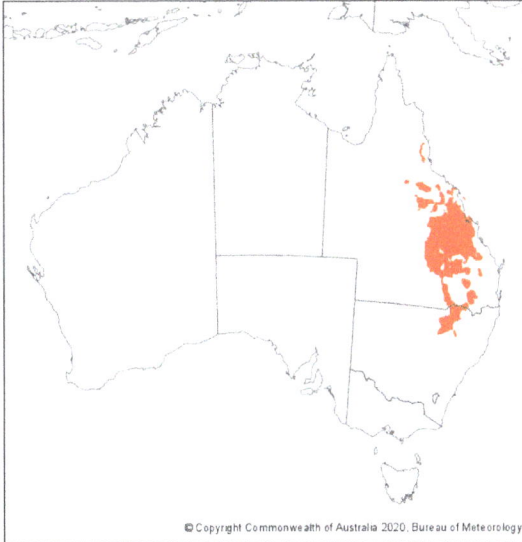

Subtropical - moderately dry winter

Species   108   Specimens   1616   Adults   1373   Larvae   243

	J	F	M	A	M	J	J	A	S	O	N	D	n.d.
Family Synlestidae													
Episynlestes albicauda	1		1								1	3	
Synlestes selysi	1				1								
Synlestes tropicus													1
Synlestes weyersii			2										
Family Lestidae													
Austrolestes analis												2	
Austrolestes aridus	3			2			1		3	1			
Austrolestes leda		3	6	2	6	1	4		4	4	22	4	1
Austrolestes psyche	1												
Indolestes tenuissimus												1	1
Lestes concinnus						1			2			1	
Family Argiolestidae													
Austroargiolestes aureus										1			
Austroargiolestes icteromelas	6	4	20	7		1			4	9	8	2	

	J	F	M	A	M	J	J	A	S	O	N	D	n.d.
Family Lestoideidae													
Diphlebia euphoeoides	1		2									3	
Diphlebia nymphoides			2									1	
Lestoidea barbarae											2		
Lestoidea conjuncta	1										2		
Family Isostictidae													
Labidiosticta vallisi	2	1		3					1				
Rhadinosticta banksi			2										
Rhadinosticta simplex	3									1	3	2	
Family Platycnemidae													
Nososticta solida	11	1	19	1					1	6	8	2	
Nososticta solitaria		1	1							1	1	3	
Family Coenagrionidae													
Aciagrion fragile												2	
Agriocnemis argentea										2			1
Agriocnemis pygmaea	1		5	1							7	1	
Agriocnemis rubricauda			3										
Argiocnemis rubescens		1	2	1							1	1	
Austroagrion watsoni	3	2		2					3	3	3	6	
Austrocnemis splendida	2			1					1		2	1	
Ceriagrion aeruginosum												1	
Ischnura aurora	1		2	6	1			1	4	5	17	14	1
Ischnura heterosticta	3	1	6	7			1	3	4	4	12	5	
Pseudagrion aureofrons	11	6	7	1						3	5	9	1
Pseudagrion cingillum	2											7	
Pseudagrion ignifer		3	1							1	2	7	
Pseudagrion microcephalum	1								2		3	1	
Xanthagrion erythroneurum	1		2	1					3	1	4	7	
Family Aeshnidae													
Adversaeschna brevistyla	6			5							3	1	
Anax guttatus	1											1	1
Austroaeschna muelleri			4	1								1	2
Austroaeschna pinheyi		46		2						10	1		
Austroaeschna speciosa			1									3	
Austrogynacantha heterogena	4	3	11	1						3	4	8	3
Dromaeschna forcipata											1	1	
Dromaeschna weiskei											1		
Gynacantha dobsoni		3		1								2	

	J	F	M	A	M	J	J	A	S	O	N	D	n.d.
Hemianax papuensis	2	2	9	5		1		2	1	1	1	4	
Spinaeschna watsoni	1											1	
Family Petaluridae													
Petalura ingentissima		1											
Family Gomphidae													
Antipodogomphus acolythus	1	2	1							1	1		
Antipodogomphus proselythus	1									1			
Austroepigomphus praeruptus	2	2								1	2	7	
Austroepigomphus turneri		1								2		1	
Austrogomphus amphiclitus		2	3	1						3	2	7	3
Austrogomphus arbustorum	2										1		
Austrogomphus australis	2										4		
Austrogomphus bifurcatus			4									3	
Austrogomphus cornutus			2							12	4	1	2
Austrogomphus divaricatus			2							1	4	4	
Hemigomphus comitatus												1	
Hemigomphus heteroclytus	1		1	2						2		3	
Ictinogomphus australis	7	3	3	1						2	1	12	2
Family Synthemistidae													
Choristhemis flavoterminata	1	3	2	1							6	4	
Eusynthemis deniseae										1		1	
Eusynthemis nigra										5	21	14	
Parasynthemis regina	3		1		2					1	1		
Tonyosynthemis claviculata			1								3		
Family Macromiidae													
Macromia tillyardi	1									2	2		
Family Corduliidae													
Hemicordulia australiae		2								3	1	2	
Hemicordulia continentalis											1		
Hemicordulia intermedia	2	2	5	1						1	4	4	
Hemicordulia tau		1	2	7						2	3	3	
Family Libellulidae													
Aethriamanta circumsignata	1										1		
Aethriamanta nymphaea										1			
Brachydiplax denticauda	1	1									1		
Crocothemis nigrifrons	5	1	4	2	1		2	1		2	1		
Diplacodes bipunctata	5	1	3	8	2		1		8	9	17	13	
Diplacodes haematodes	8	5	9	5		1	1		2	15	10	7	

	J	F	M	A	M	J	J	A	S	O	N	D	n.d.
Diplacodes nebulosa												1	
Diplacodes trivialis	2	2			1						1		
Hydrobasileus brevistylus		2	1	1						1		3	
Lathrecista asiatica festa								1				2	
Macrodiplax cora	1	1											
Nannodiplax rubra		2											
Nannophlebia eludens												1	
Nannophlebia risi			1							5	6	3	1
Nannophya australis											5		1
Neurothemis stigmatizans	3							2				1	
Orthetrum caledonicum	4	2	9	18		1			2	3	14	9	
Orthetrum migratum	6									1	1		
Orthetrum sabina	1	3			1					1		2	
Orthetrum villosovittatum	1	2	1	13					1	2	14	3	
Pantala flavescens	2	1	5	1							1		
Potamarcha congener		20									1		
Rhodothemis lieftincki	3	3	2									5	
Rhyothemis braganza	2									1			
Rhyothemis graphiptera	8	2	2					4		2	2	2	1
Rhyothemis phyllis	1	1											1
Rhyothemis princeps			1									1	
Tholymis tillarga		2	3										
Tramea loewii	3	4	4	1					6	3	2	3	
Zyxomma elgneri	2	2	6	3						1	2	4	
Genera incertae sedis													
Austrocordulia refracta											1	2	
Austrophya mystica												1	
Cordulephya bidens				1									
Micromidia·atrifrons	1	3	3	1							3		
Pseudocordulia elliptica												1	

Subtropical - Distinctly dry winter

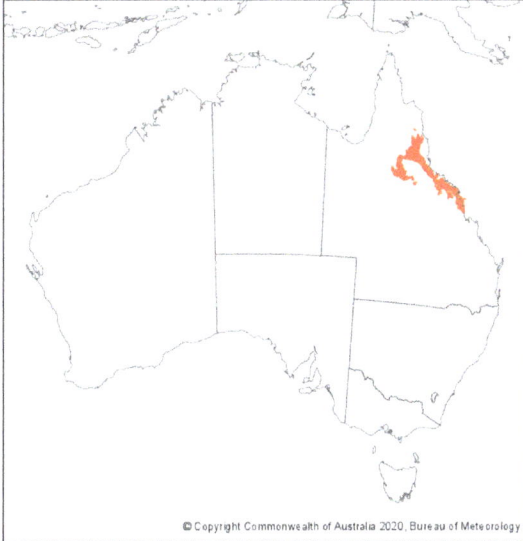

Subtropical - distinctly dry winter

Species   106   Specimens   1111   Adults   1111   Larvae   0

	J	F	M	A	M	J	J	A	S	O	N	D	n.d.
Family Synlestidae													
Chorismagrion risi		1	1										1
Episynlestes cristatus											1		
Synlestes selysi												1	
Family Lestidae													
Austrolestes insularis			3		3				1				
Austrolestes leda	6								1				
Indolestes tenuissimus											1		
Lestes concinnus			1	1		1			2	1		1	
Family Argiolestidae													
Austroargiolestes aureus			3										
Austroargiolestes icteromelas	4	1	2						3		1	4	
Family Lestoideidae													
Diphlebia coerulescens	1												
Diphlebia euphoeoides	1	1	12	6	1						5	5	

	J	F	M	A	M	J	J	A	S	O	N	D	n.d.
Lestoidea conjuncta		1									2		
Family Isostictidae													
Labidiosticta vallisi			1	3									
Neosticta fraseri		1	10								1		
Oristicta filicicola				1						1			
Rhadinosticta banksi			1										
Rhadinosticta simplex			1								2		
Family Platycnemidae													
Nososticta coelestina	1		1								1	1	
Nososticta solida	2		7	2					8	3	2	6	
Nososticta solitaria	2	6	17	2	2				2	3	12	16	
Family Coenagrionidae													
Aciagrion fragile											1	1	
Agriocnemis argentea			1	3	2					4	1	12	
Agriocnemis pygmaea	1			2	1				3	1	3	3	
Agriocnemis rubricauda										1			
Argiocnemis rubescens			1			1	1		10	1	1	1	
Austroagrion exclamationis	1	2		1						1	3	1	
Austroagrion watsoni			1	1					7	1	2	10	
Austrocnemis splendida			1								4	1	
Ceriagrion aeruginosum		1			1						1	1	
Ischnura aurora		1	4		1				5	4	1	5	
Ischnura heterosticta	3	1	3	3	1	1	1		4	1	4	5	
Ischnura pruinescens		1	1	2					3		5	1	
Pseudagrion aureofrons			1						1	1	1	1	
Pseudagrion ignifer	9	1	2	1	2			1	6	3	11	14	
Pseudagrion microcephalum		1	3	3	1				4	3	8	5	1
Xanthagrion erythroneurum									1				
Family Aeshnidae													
Adversaeschna brevistyla											1		
Anax gibbosulus			1										
Anax guttatus		1											
Austroaeschna speciosa	2	2	1								1		
Austrogynacantha heterogena			2						1				
Dromaeschna forcipata	4	4									6		
Dromaeschna weiskei	1										2		
Gynacantha dobsoni										1	2		
Gynacantha nourlangie								1			1		

	J	F	M	A	M	J	J	A	S	O	N	D	n.d.
Gynacantha rosenbergi										1	2		
Hemianax papuensis	3									1		1	
Spinaeschna watsoni	1	2											
Telephlebia tillyardi	1		1								6	1	
Family Petaluridae													
Petalura ingentissima	2												
Family Gomphidae													
Antipodogomphus proselythus													1
Austroepigomphus turneri											3	1	
Austrogomphus amphiclitus	1		1	2						2	2	3	
Austrogomphus arbustorum											1		
Austrogomphus bifurcatus			2										
Austrogomphus cornutus												3	
Austrogomphus divaricatus	1	3									1		
Austrogomphus doddi	2	2		1									
Austrogomphus prasinus	2	2									4	1	
Hemigomphus comitatus	4	1	1	1						1	5	3	
Hemigomphus heteroclytus											2		
Ictinogomphus australis			2							1	4	9	1
Family Synthemistidae													
Choristhemis flavoterminata	5	2									6	7	
Eusynthemis nigra	2	1	2							2	1	1	
Parasynthemis regina											1		
Tonyosynthemis claviculata	1	4		1									
Family Macromiidae													
Macromia tillyardi	2										1		
Family Corduliidae													
Hemicordulia australiae										1	1	1	
Hemicordulia intermedia	4		1	1							5	1	
Pentathemis membranulata											4	1	
Family Libellulidae													
Aethriamanta circumsignata												2	
Agrionoptera l.biserialis											1		
Brachydiplax denticauda			1	1						4	7	1	
Crocothemis nigrifrons	2		2	3	1					3	3	6	
Diplacodes bipunctata	8		5	4	2			12	13	8	17	2	
Diplacodes haematodes	5	1	4	6	12	1	1	1	7	4	3	9	
Diplacodes nebulosa		1		1							1		

	J	F	M	A	M	J	J	A	S	O	N	D	n.d.
Diplacodes trivialis		1	3	1	3			8	6	4	15	1	1
Hydrobasileus brevistylus	1	4	3	1						1	4	1	
Lathrecista asiatica festa			2								2		
Macrodiplax cora		1									3	1	
Nannodiplax rubra	1		1		3	1			3	3	2	3	
Nannophlebia eludens	1	3		1							1	1	
Nannophlebia risi	1	1	4	1	2					2	2	2	
Neurothemis stigmatizans	1	2	2	2	6				2	1	7		1
Orthetrum caledonicum	6	1	1	2	2			3	11	3	3	3	
Orthetrum migratum			1	1							2	1	
Orthetrum sabina			1	2	2			1		4	1	1	1
Orthetrum serapia											1		
Orthetrum villosovittatum	4	1	2	2	3				3		2	2	
Pantala flavescens			1	1					1	2	1		
Rhodothemis lieftincki				1						1	5	4	
Rhyothemis braganza	1	1								1	6	10	
Rhyothemis graphiptera	1		1	2					1	2	7	2	1
Rhyothemis phyllis			2	3						3	4	1	
Rhyothemis princeps											2		1
Tetrathemis i.cladophila											2	4	
Tholymis tillarga			2								3		
Tramea eurybia		1											
Tramea loewii	3		3	4	3	1		1	1	1	2	1	
Urothemis aliena											1		
Zyxomma petiolatum											1		
Genera incertae sedis													
Archaeophya magnifica		2											
Austrophya mystica											2	1	
Micromidia·atrifrons			2	1									
Pseudocordulia elliptica											1		

Equatorial – Savanna

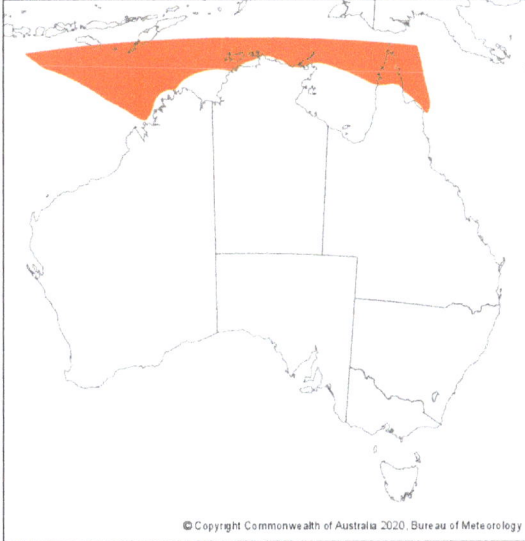

Equatorial - savanna

Species   104   Specimens   2423   Adults   2423   Larvae   0

	J	F	M	A	M	J	J	A	S	O	N	D	n.d.
Family Lestidae													
Austrolestes insularis	1	14	24		1		4	3					
Indolestes alleni	2				1	2	2	1		1	1	3	
Indolestes tenuissimus	4	2								2		5	
Lestes concinnus	2		7				1					2	
Family Argiolestidae													
Podopteryx selysi													1
Family Isostictidae													
Oristicta filicicola			6	14	1	1				1	9		
Rhadinosticta banksi		3											
Family Platycnemidae													
Nososticta coelestina	3		35			10	1	3	4		6	6	
Nososticta fraterna										32		12	
Nososticta koolpinyah			1			13		1		60			
Nososticta solida		1	1										

283

	J	F	M	A	M	J	J	A	S	O	N	D	n.d.
Nososticta solitaria	3		8			1	3		1			4	
Nososticta taracumbi	1					10	1			21			
Family Coenagrionidae													
Aciagrion fragile	2	1	5	4		1	2	1	5	24	19		
Agriocnemis femina				2									
Agriocnemis pygmaea	1		2	2	1		1		1	3	2		
Agriocnemis rubricauda									31	19			
Archibasis mimetes	3					1	1						
Argiocnemis rubescens						1		14	21		1		
Austroagrion exclamationis		3	8	2	5	1	4	5		31	7	5	
Austrocnemis maccullochi									25				
Ceriagrion aeruginosum	3			1	2	1	11	6	16	1	2	2	
Ischnura aurora			2		5	2			1		2		
Ischnura heterosticta	3	10	5	16	1	4	8	1	4	16		2	
Ischnura pruinescens				1					1	8			
Pseudagrion cingillum				1							1		
Pseudagrion ignifer				4					3	9		1	
Pseudagrion lucifer			16			1	4	2				5	
Pseudagrion microcephalum	2		2		14		2	1	2	11			
Teinobasis rufithorax	11	4	8	22		3	4				1	4	
Family Aeshnidae													
Agyrtacantha dirupta	1					3	3				1		
Anaciaeschna jaspidea	1	2	2		2								1
Anax gibbosulus	2	1	1				1	2		2	1	1	
Anax guttatus	8	5	6	5	1				1		1	13	
Austrogynacantha heterogena	1	1				1							
Gynacantha dobsoni	1		1	1		21	11	1		3			
Gynacantha kirbyi							1	1			1		
Gynacantha mocsaryi	2		1	2		4				1	1	3	
Gynacantha rosenbergi	3	3				1					7		
Hemianax papuensis		1	7	1		1		1					
Family Petaluridae													
Petalura pulcherrima	1											1	
Family Gomphidae													
Antipodogomphus edentulus	1												
Antipodogomphus proselythus	1											4	
Austroepigomphus turneri	1									1	2		
Austrogomphus arbustorum										1			

	J	F	M	A	M	J	J	A	S	O	N	D	n.d.
Austrogomphus mjobergi	2		2	1									
Austrogomphus prasinus				5	1							3	
Ictinogomphus australis	1		1							1	3	1	
Ictinogomphus paulini	2									2	1		
Family Macromiidae													
Macromia tillyardi			3							2			
Macromia viridescens	1		1	1									1
Family Corduliidae													
Hemicordulia continentalis	1												
Hemicordulia kalliste	2												
Metaphya tillyardi												1	1
Pentathemis membranulata	3									4		1	
Family Libellulidae													
Aethriamanta circumsignata	1									1			
Aethriamanta nymphaea	1		1										
Agrionoptera i.allogenes	2	4	4	3		1	1	2			1	3	
Agrionoptera l.biserialis	2			1				1					
Brachydiplax denticauda	5	1		5		1		1	8	3		2	3
Brachydiplax duivenbodei	2										1		
Camacinia othello	2	3			1			1	1				2
Crocothemis nigrifrons	1						1						1
Diplacodes bipunctata	3	8	12	6	2	2				1	2		1
Diplacodes haematodes	5	1	9	2	9	4	16	3	11	6	2	3	
Diplacodes nebulosa	2		11			1		2			4		
Diplacodes trivialis	7	13	5	5	2	6	5		3	4	1	1	2
Huonia melvillensis										5			
Hydrobasileus brevistylus			12		2		4	3				2	2
Lathrecista asiatica festa	15	9	11	2	6	7	4	6	18	3	4	8	
Macrodiplax cora	2	2	1	1	2		2		1	2	2	3	
Nannodiplax rubra	7		18	3		2	9	28	4	22	1	3	3
Nannophlebia eludens		1	7	4		3	1		1	1			
Nannophlebia mudginberri									3				
Nannophlebia risi	1		1					1					
Nannophya australis	1	7						1					
Nannophya paulson	2												
Neurothemis oligoneura	3	9		1	1	12	4	9	14	3	1	3	5
Neurothemis stigmatizans	14	17	21	8	9	37	25	13	38	13	5	18	9
Orthetrum balteatum	2						1			1			

	J	F	M	A	M	J	J	A	S	O	N	D	n.d.
Orthetrum caledonicum	2		3				2	2	1	4	1	4	1
Orthetrum migratum	5		18						1	5			
Orthetrum sabina	7	3	6	2	2	1	1	1	9	2	1	5	3
Orthetrum serapia	2			2	2	4	6	3	2		5		3
Orthetrum villosovittatum	6		2	2		3	1	2	4	1	1	4	2
Pantala flavescens	9	17	11	1	1	3					1	1	1
Potamarcha congener	1		1								2		
Raphismia bispina					4								
Rhodothemis lieftincki	4	1					1	1			3		
Rhyothemis braganza	2	3		1					2	2	5	1	1
Rhyothemis graphiptera	10		7	4	3	6			2		1	2	2
Rhyothemis phyllis	7	1	4							1	2	5	1
Rhyothemis princeps	2	1											
Rhyothemis resplendens	8	1	1									8	
Tetrathemis irregularis		2	1			1	1			1	1	2	
Tholymis tillarga	5	3	6	3	2	12	5	3	1	1	3	9	
Tramea loewii	6	8	33	3	8	5	6	2	1	1	1	1	3
Tramea propinqua			2				1						
Tramea stenoloba	1	2								1		2	
Urothemis aliena	5									1			
Zyxomma elgneri							1					4	
Zyxomma multinervorum	2												
Zyxomma petiolatum											1	2	
Genera incertae sedis													
Micromidia rodericki	1												

Tropical - Rainforest (persistently wet)

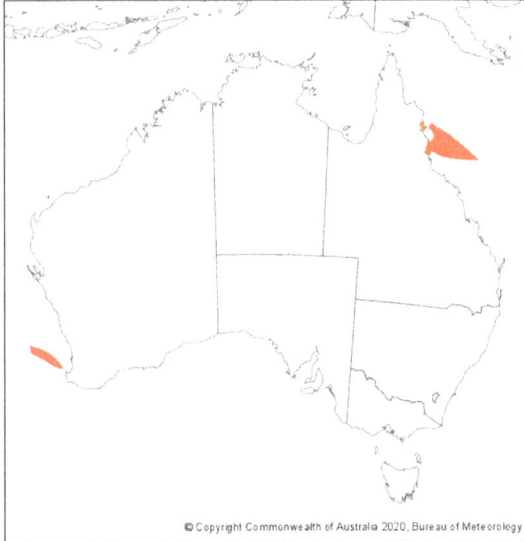

Tropical - rainforest (persistently wet)

Species 103 Specimens 1071 Adults 1071 Larvae 0

	J	F	M	A	M	J	J	A	S	O	N	D	n.d.
Family Synlestidae													
Chorismagrion risi	1		2								2		
Episynlestes cristatus											2		
Synlestes tropicus			1								5		
Family Lestidae													
Indolestes alleni				1									
Indolestes tenuissimus			1								2	2	
Lestes concinnus					1								
Family Argiolestidae													
Austroargiolestes aureus	1	6	3						2	7	5	3	
Austroargiolestes icteromelas								1					
Griseargiolestes metallicus	1								5	7	2		
Podopteryx selysi	1		1	1						1			
Family Lestoideidae													
Diphlebia euphoeoides	7	7	7	1		1		6	20	6	5	13	

	J	F	M	A	M	J	J	A	S	O	N	D	n.d.
Diphlebia hybridoides										2	5		
Lestoidea barbarae											2	2	
Lestoidea brevicauda			1							4	2		
Lestoidea conjuncta	5	1		2				1	3	6			
Family Isostictidae													
Neosticta fraseri	1		1					1					
Oristicta filicicola	1	2	4					2	1	3	2	1	
Rhadinosticta simplex			2						1	1			
Family Platycnemidae													
Nososticta coelestina		1	4	2					1	4	3	12	
Nososticta solitaria	4		8	1				1	5	10	2	25	
Family Coenagrionidae													
Aciagrion fragile						1			4	4	5	1	
Agriocnemis argentea									1		1		
Agriocnemis dobsoni	3								2	2	7		
Agriocnemis pygmaea									1		2		
Archibasis mimetes				4					7	7	2		
Argiocnemis rubescens			1	1				1	4	1	2	3	
Austroagrion exclamationis											1		
Ceriagrion aeruginosum	1			1					10	1	3	2	
Ischnura aurora	1	1			1						1	4	
Ischnura heterosticta			1					1	1	1	1		
Ischnura pruinescens				3									
Pseudagrion aureofrons											1		
Pseudagrion ignifer	2		1	1				3	16	4	3	4	
Pseudagrion microcephalum	1		2						9	2	2	2	
Teinobasis rufithorax			3						1	7		2	
Family Aeshnidae													
Anax gibbosulus					1				1				
Anax guttatus		1											
Austrogynacantha heterogena		1	1										
Austrophlebia subcostalis											1	1	
Dromaeschna forcipata	4		1							3	4	1	1
Dromaeschna weiskei					1					1	2	1	1
Gynacantha dobsoni	1							2	1	6	1		
Gynacantha kirbyi										1			
Gynacantha mocsaryi		1	1						1	3			
Gynacantha rosenbergi	1		1		1				5	1	2		

	J	F	M	A	M	J	J	A	S	O	N	D	n.d.
Hemianax papuensis	2	3	1								1		
Spinaeschna watsoni									1		1		
Telephlebia tillyardi			1								1	1	
Family Petaluridae													
Petalura ingentissima	3		1								1	3	1
Family Gomphidae													
Antipodogomphus acolythus											1		
Austrogomphus amphiclitus											1		
Austrogomphus bifurcatus	1			1					1	1	1	1	
Austrogomphus divaricatus										2			
Austrogomphus doddi									1				
Austrogomphus prasinus	3								1	2	7	6	
Hemigomphus comitatus	2									1	4	2	
Hemigomphus theischingeri			1								2	3	
Ictinogomphus australis		1									1		
Odontogomphus donnellyi											1		
Family Synthemistidae													
Choristhemis flavoterminata	1		1						3	4	4	10	
Choristhemis olivei	1												
Eusynthemis nigra										1	6	1	3
Family Corduliidae													
Hemicordulia australiae			1						1		1		
Hemicordulia continentalis	1	1	1						3	3	3	2	
Hemicordulia intermedia			1								1		
Family Libellulidae													
Aethriamanta circumsignata									2			1	
Agrionoptera i.allogenes	1	2	6	8		2			1	5	2	1	
Agrionoptera l.biserialis		1	2								4	2	
Brachydiplax denticauda			2	2							2		
Brachydiplax duivenbodei	9		13						1				
Camacinia othello		1											
Crocothemis nigrifrons									1				
Diplacodes bipunctata			2			1	1			1			
Diplacodes haematodes					1	3		1	5		1		
Diplacodes nebulosa			1						1				
Diplacodes trivialis		1	1	2							2	1	
Hydrobasileus brevistylus	1									2	4	1	
Lathrecista asiatica festa				2				3	2	2			

	J	F	M	A	M	J	J	A	S	O	N	D	n.d.
Macrodiplax cora	1												
Nannodiplax rubra	3		4	1		5		1	4	4	4	1	
Nannophlebia eludens	1		1							1	1	8	
Nannophlebia risi											1		
Neurothemis stigmatizans	6	2	20		2	4	1	5	12	8	3	14	1
Orthetrum caledonicum					1	4							
Orthetrum sabina		1	1	3	1	7			2	3	4	3	
Orthetrum serapia									1	1			
Orthetrum villosovittatum	7		5			6	1	1	6	18	4	6	
Pantala flavescens		1			1								
Potamarcha congener			1										
Rhodothemis lieftincki											1	1	
Rhyothemis phyllis	4		2	1					7	2	1		
Rhyothemis princeps	7		1	3				2	2	6	7		
Rhyothemis resplendens	9	1							2		4	3	
Tetrathemis i. cladophila		3	3	3					1		2	4	
Tholymis tillarga	1				1				2		1		
Tramea loewii		2	2	1	1			3	2	1	1	1	
Tramea propinqua											1	1	
Urothemis aliena											1		
Zyxomma petiolatum										1		1	
Genera incertae sedis													
Archaeophya magnifica											1	1	2
Austrophya mystica			1							1			
Pseudocordulia circularis									1	7	1		
Pseudocordulia elliptica										2			

Temperate - No dry season (mild Summer)

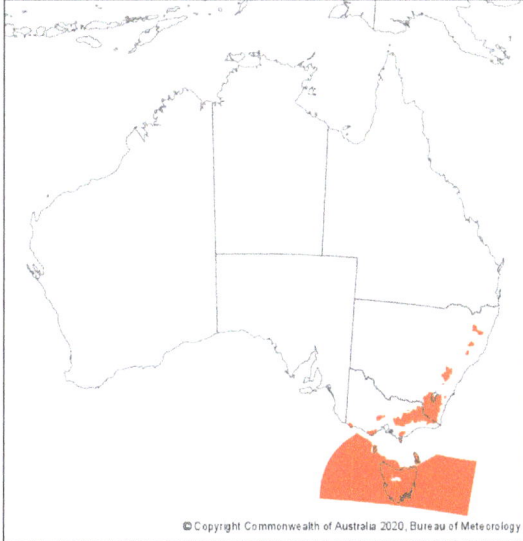

Temperate - no dry season (mild summer)

Species   95   Specimens   4788   Adults   4475   Larvae   313

	J	F	M	A	M	J	J	A	S	O	N	D	n.d.
Family Hemiphlebiidae													
Hemiphlebia mirabilis	10	1		1							5	4	1
Family Synlestidae													
Synlestes selysi	1												
Synlestes weyersii	47	87	15	10	2						4	41	1
Family Lestidae													
Austrolestes analis	28	21	4		1					5	20	20	1
Austrolestes annulosus	25	13	11	2	5	5	7		6	17	26	16	1
Austrolestes aridus												1	
Austrolestes cingulatus	20	48	3			1				3	18	42	2
Austrolestes io	3	3	3					2	6	3	2		
Austrolestes leda	12	12	2							6	6	28	
Austrolestes minjerriba		1	1										
Austrolestes psyche	24	42	12				4		3	11	16	26	2
Family Argiolestidae													

	J	F	M	A	M	J	J	A	S	O	N	D	n.d.
Austroargiolestes alpinus	12	2								1	1	1	1
Austroargiolestes brookhousei	4											4	4
Austroargiolestes calcaris	52	14	3							2	5	63	1
Austroargiolestes christine	1											6	26
Austroargiolestes icteromelas	35	25	2	1					1	10	17	93	2
Austroargiolestes isabellae	3										2	20	
Griseargiolestes bucki		1										3	1
Griseargiolestes eboracus	12	6								3	10	26	13
Griseargiolestes griseus	35	1									6	17	1
Griseargiolestes intermedius	8	34	2							1	1	9	
Family Lestoideidae													
Diphlebia lestoides	11	1								5	15	57	2
Diphlebia nymphoides	5	1									2	7	
Family Isostictidae													
Neosticta canescens											1	2	
Rhadinosticta simplex		1										1	
Family Platycnemidae													
Nososticta solida												1	
Family Coenagrionidae													
Agriocnemis pygmaea											1	1	
Austroagrion watsoni	27	19	4						4	4	6	12	1
Austrocnemis splendida												1	
Ausrocoenagrion lyelli	10	4								2	9	7	2
Ischnura aurora	15	15	7						2	5	19	24	
Ischnura heterosticta	29	13	4	2	1	2	2		8	13	25	24	3
Pseudagrion ignifer	1												
Pseudagrion microcephalum													1
Xanthagrion erythroneurum	7	10	5		1					3	9	10	
Family Austropetaliidae													
Archipetalia auriculata		1											1
Austropetalia patricia										1	6	2	2
Austropetalia tonyana										4	13		1
Family Aeshnidae													
Acanthaeschna victoria											2		
Adversaeschna brevistyla	20	23	12	6	2	5			3	7	14	29	4
Austroaeschn atrata	13	6	6									3	2
Austroaeschna flavomaculata	9	22	5								1		
Austroaeschna hardyi	15	37	17	2	2				1	12	1	1	

	J	F	M	A	M	J	J	A	S	O	N	D	n.d.
Austroaeschna inermis	6	12	10	1							2		1
Austroaeschna multipunctata	13	49	17								1	2	32
Austroaeschna obscura	6	4	1	1								3	1
Austroaeschna parvistigma	23	31	20	2	1					3	6	13	5
Austroaeschna pulchra	22	10	12	3						1	2	6	3
Austroaeschna sigma	8	5	3								2	5	2
Austroaeschna subapicalis	16	17	5	2							2	7	8
Austroaeschna tasmanica	4	13	2	1						6			
Austroaeschna unicornis	11	11	11	1					5	10	2	3	3
Austrophlebia costalis	1												
Hemianax papuensis	9	6	4	1						3	2	5	1
Notoaeschna geminata	4	1								3	3	5	
Notoaeschna sagittata	13	4	1							1	3	13	1
Spinaeschna tripunctata	8	1										1	
Telephlebia brevicauda	38	25	8	2		1			1		3	9	2
Telephlebia godeffroyi	12		3								3	3	1
Family Petaluridae													
Petalura gigantea	76	12	1			1					1	33	
Family Gomphidae													
Austrogomphus amphiclitus			1							1		1	
Austrogomphus cornutus	1											2	
Austrogomphus guerini	59	47	7	1					5	3	10	44	4
Austrogomphus ochraceus	4	8	2								3	5	
Hemigomphus gouldii	20	5							1		2	20	2
Hemigomphus heteroclytus	3											2	
Family Synthemistidae													
Archaeosynthemis orientalis	15	4	2								1	1	1
Choristhemis flavoterminata	1												
Eusynthemis aurolineata	3	10									1	6	14
Eusynthemis brevistyla	42	14									4	44	9
Eusynthemis guttata	26	35	13	1							1	34	
Eusynthemis rentziana	2											3	
Eusynthemis tillyardi	13	2									3	18	1
Eusynthemis ursa												1	
Eusynthemis ursula												1	
Eusynthemis virgula	16	3	1								1	10	2
Syn. gomphomacromioides	5	28	6							1			
Synthemis eustalacta	58	71	13							1	10	42	5

	J	F	M	A	M	J	J	A	S	O	N	D	n.d.
Synthemis tasmanica	18	42	9		1							2	2
Family Corduliidae													
Hemicordulia australiae	1	1	1								3	4	
Hemicordulia intermedia	1												
Hemicordulia tau	23	34	8						2	8	15	32	
Procordulia jacksoniensis	25	24	6		1		1		4	8	18	16	3
Family Libellulidae													
Austrothemis nigrescens	3	2									1		
Diplacodes bipunctata	23	16	1	1							8	17	
Diplacodes haematodes	1	1	2	2							4	7	2
Diplacodes melanopsis		2									1		
Nannophya australis	3	3	1							1	1		2
Nannophya dalei	28	40	1							5	10	16	2
Orthetrum caledonicum	6	5	2								2	5	2
Orthetrum villosovittatum	1	2									1		
Rhyothemis graphiptera		1											
Genera incertae sedis													
Cordulephya montana	10	1										2	

Grassland - Hot (Winter drought)

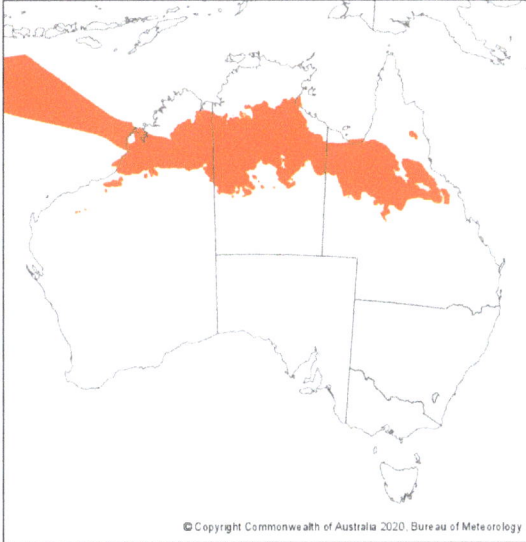

Grassland - hot (winter drought)

Species 90   Specimens 1421   Adults 1419   Larvae 2

	J	F	M	A	M	J	J	A	S	O	N	D	n.d.
Family Lestidae													
Austrolestes annulosus	1												
Austrolestes aridus		1	1	2				3		2			1
Austrolestes insularis				1			1	4			1		
Austrolestes leda				3							1		
Lestes concinnus	1	5		6	5		3	1			4	1	
Family Argiolestidae													
Austroargiolestes icteromelas				12							1		
Family Isostictidae													
Austrosticta fieldi			1	6						1			
Austrosticta frater				16									1
Austrosticta soror			11										
Eurysticta kununurra	2	2											
Eurysticta reevesi				2									
Rhadinosticta banksi		4	9	5									

	J	F	M	A	M	J	J	A	S	O	N	D	n.d.
Family Platycnemidae													
Nososticta coelestina			1										
Nososticta fraterna	1			18	1	1	1		1	9	4	1	
Nososticta kalumburu	2							1				1	
Nososticta liveringa		56			12		5	14	1		4		
Nososticta solida			1							4		1	
Nososticta solitaria		4		2							1	6	
Family Coenagrionidae													
Agriocnemis argentea		2		14						5			
Agriocnemis pygmaea		2	4	2		2	1		2	4	3		
Argiocnemis rubescens				6		1				5	4	1	
Austroagrion exclamationis				2	1			1	3	3	6	1	1
Austroagrion watsoni				22	4		2	2	1	6		2	
Austrocnemis maccullochi				1						4	6	1	
Austrocnemis splendida												2	1
Ceriagrion aeruginosum	1	3				3	5			1		2	
Ischnura aurora		2	4	29	3		1	1	5	4		3	
Ischnura heterosticta		4		1	3	3	2		3	3	3	5	1
Ischnura pruinescens		1		1									1
Pseudagrion aureofrons		11	15				6	4	1	3			
Pseudagrion cingillum		4					2	4	1	8	1		
Pseudagrion ignifer			1	1								1	
Pseudagrion jedda	1			4		1	3			8	2		
Pseudagrion lucifer	1												
Pseudagrion microcephalum	1	19	11	10	1		3	3	8	3	10	1	
Xanthagrion erythroneurum	1	1	1	5				2		1	4	1	
Family Aeshnidae													
Adversaeschna brevistyla										1	1		
Anax gibbosulus		2										3	
Anax guttatus	1												
Austrogynacantha heterogena	1	7		20	1								
Gynacantha nourlangie				3	4				1	5	2	1	
Hemianax papuensis	13	5	4	6	4	1						3	2
Family Gomphidae													
Antipodogomphus acolythus				1									
Antipodogomphus neophytus	3	1		1									
Antipodogomphus proselythus												2	
Austroepigomphus turneri											2	5	

	J	F	M	A	M	J	J	A	S	O	N	D	n.d.
Austrogomphus amphiclitus													1
Austrogomphus arbustorum			1									4	
Austrogomphus cornutus										1			
Austrogomphus doddi			1										
Austrogomphus mjobergi		4			1					1		2	
Austrogomphus pusillus													1
Ictinogomphus australis		3		4	1					2	2	7	2
Family Synthemistidae													
Choristhemis flavoterminata				6									
Family Corduliidae													
Hemicordulia australiae		1							1	2			
Hemicordulia continentalis		5									1		
Hemicordulia intermedia	1			4			1		2	8			
Hemicordulia tau		1		2									
Pentathemis membranulata												2	
Family Libellulidae													
Aethriamanta circumsignata										1		4	1
Brachydiplax denticauda					1								
Crocothemis nigrifrons	1	7		1		1	2		1	2	1	8	
Diplacodes bipunctata	5	10	9	21	7	4	3	1	2	3	1	5	2
Diplacodes haematodes	5	1	6	42	1	7	7		3	12	4	19	
Diplacodes trivialis	1	8		1	1	5	1	1			1	6	
Hydrobasileus brevistylus	1	3										6	1
Lathrecista asiatica festa										1			
Macrodiplax cora		2			1					2	2		
Nannodiplax rubra	4			12	4					2	1		1
Nannophlebia injibandi				1						6	1		
Nannophlebia risi			1							1			
Neurothemis oligoneura						6				1			
Neurothemis stigmatizans		1		1				1		1	1		
Notolibellula bicolor	1				2					1	1		
Orthetrum caledonicum	6	4	8	17	1	4	1	1	2	11		7	2
Orthetrum migratum			1					1		4			1
Orthetrum sabina		4	2			3						2	
Orthetrum serapia		1											
Orthetrum villosovittatum		5		4								5	
Pantala flavescens	1	7	6	4	2	5	1		2			2	2
Potamarcha congener		3						1		1	1		

The Distribution of Australian Dragonflies

	J	F	M	A	M	J	J	A	S	O	N	D	n.d.
Rhodothemis lieftincki		1		1							2	2	
Rhyothemis braganza	3	1	1	4						2		2	
Rhyothemis graphiptera	4	3	3	3	2	1	1			4	5	4	
Rhyothemis phyllis			1										
Tholymis tillarga		2				1				2			
Tramea loewii	2	8		11	1	2	2	2	4	4		5	2
Tramea propinqua													
Tramea stenoloba		1	1	2									
Zyxomma elgneri			1	8						1	1	1	

Temperate - Distinctly dry (and warm) Summer

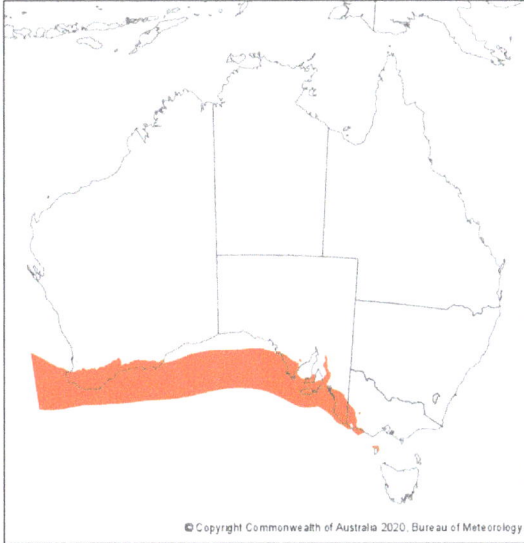

Temperate - distinctly dry (and warm) summer

Species   54   Specimens   1387   Adults   1380   Larvae   7

	J	F	M	A	M	J	J	A	S	O	N	D	n.d.
Family Hemiphlebiidae													
Hemiphlebia mirabilis	6		1								4	17	16
Family Lestidae													
Austrolestes aleison		4		2								3	2
Austrolestes analis	20	5	10	10				1		1	27	42	3
Austrolestes annulosus	14		2	7						17	17	24	2
Austrolestes aridus											1		
Austrolestes cingulatus								1					
Austrolestes io	5	3	1				1			2	8	8	1
Austrolestes leda			1									13	1
Austrolestes psyche	1		1							6	6	14	
Family Argiolestidae													
Archiargiolestes parvulus										2		1	
Archiargiolestes pusillissimus	1	1								4	30	7	

	J	F	M	A	M	J	J	A	S	O	N	D	n.d.
Archiargiolestes pusillus									1	49	49	41	2
Austroargiolestes icteromelas												1	
Miniargiolestes minimus	17		2							5	12	16	
Family Coenagrionidae													
Austroagrion cyane	1			3						2	13	19	
Austroagrion watsoni											3	12	
Ausrocoenagrion lyelli											1	1	
Ischnura aurora		2	4	2				1	2	8	11	32	
Ischnura heterosticta	7	3	5	1						6	12	29	5
Xanthagrion erythroneurum	4	1	3	2					1	1	5	16	
Family Aeshnidae													
Adversaeschna brevistyla	8	5	2	3				2	2	19	28	2	
Austroaeschna anacantha	4	1	1								7	3	
Austroaeschna parvistigma		1	1	1							4	2	
Austroaeschna unicornis				1		1					1		
Hemianax papuensis	6	3	9	4			1	1	4	2	13	3	
Family Petaluridae													
Petalura hesperia											1		
Family Gomphidae													
Armagomphus armiger	1									1	1		
Austroepigomphus praeruptus												1	
Austrogomphus australis	1												
Austrogomphus collaris	8										29	1	
Austrogomphus cornutus				1									
Austrogomphus guerini		1								1	8	1	
Zephyrogomphus lateralis	2									2	10		
Family Synthemistidae													
Archaeosynthemis leachii	5	1	1						1	2	9		
Archaeosynthemis occidentalis	5									2			
Austrosynthemis·cyanitincta										2	3		
Eusynthemis brevistyla										2			
Synthemis eustalacta	2		1						3	16	1		
Family Corduliidae													
Hemicordulia australiae	5		1							5			
Hemicordulia tau	6	2	6	15				2	13	26	20		
Procordulia affinis	2								3		3	2	
Procordulia jacksoniensis	2								3	13	9	2	
Family Libellulidae													

	J	F	M	A	M	J	J	A	S	O	N	D	n.d.
Austrothemis nigrescens	7										7	10	8
Crocothemis nigrifrons	2		3									4	
Diplacodes bipunctata	4	1	5	3				2	8	18	6	28	1
Diplacodes haematodes		1		1						1	2	3	3
Diplacodes melanopsis											1	5	
Nannophya dalei	1										2	2	
Nannophya occidentalis												2	
Orthetrum caledonicum	5	4	1	1							2	13	1
Pantala flavescens		1	1	2									
Tramea loewii										1			
Genera incertae sedis													
Hesperocordulia berthoudi	1			1								2	
Lathrocordulia metallica											1	1	

Grassland - Hot (persistently dry)

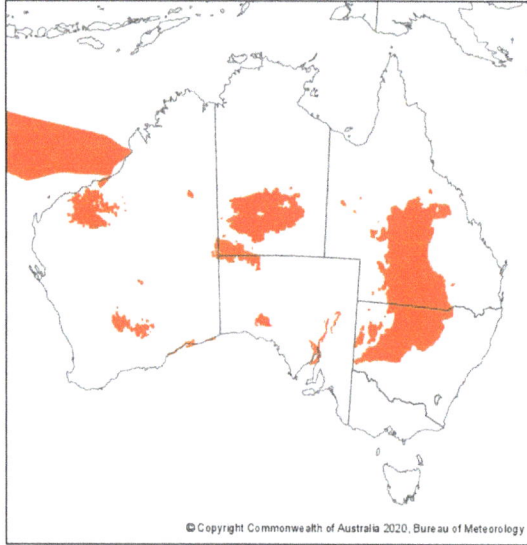

© Copyright Commonwealth of Australia 2020, Bureau of Meteorology

Grassland - hot (persistently dry)

Species 52   Specimens 1463   Adults 939   Larvae 524

	J	F	M	A	M	J	J	A	S	O	N	D	n.d.
Family Lestidae													
Austrolestes annulosus	5		1						1				
Austrolestes aridus	1	1	6		2			2	6	6	4	2	
Austrolestes insularis											1		
Austrolestes leda	1		3		1	1			2	3	3		
Lestes concinnus										1			
Family Isostictidae													
Rhadinosticta simplex										2			
Family Platycnemidae													
Nososticta liveringa		1									2		
Nososticta solida										2			
Family Coenagrionidae													
Agriocnemis argentea	1				1			2			2		
Agriocnemis kunjina		3					1	2			3		
Argiocnemis rubescens							1			1			

	J	F	M	A	M	J	J	A	S	O	N	D	n.d.
Austroagrion pindrina								11			3		
Austroagrion watsoni					1				2	2	1		
Ischnura aurora	5	5	46	5	30	1	1	2	16	3	9	5	1
Ischnura heterosticta	1	4	3		1		1	2	3	2	1	1	
Pseudagrion aureofrons								15	3	7	2		
Pseudagrion microcephalum					2					1	2		
Xanthagrion erythroneurum	12	22	6	5	7		1	1	23	6	1	7	2
Family Aeshnidae													
Austrogynacantha heterogena	3		2	1						2	2	1	
Hemianax papuensis	5	6	13	10	7		2	3	5		2	3	2
Family Gomphidae													
Antipodogomphus acolythus										1			
Austroepigomphus gordoni											3		
Austrogomphus amphiclitus											1		
Austrogomphus australis										10		1	
Austrogomphus mjobergi											2		
Austrogomphus ochraceus											1		
Ictinogomphus dobsoni											7		
Family Synthemistidae													
Choristhemis flavoterminata											1		
Eusynthemis nigra													1
Family Corduliidae													
Hemicordulia flava	2											3	
Hemicordulia intermedia					1								
Hemicordulia koomina							3			2			
Hemicordulia tau	2	3	4	3	7		1	3	11	1	1		
Family Libellulidae													
Crocothemis nigrifrons			3	2						1	1		
Diplacodes bipunctata		2	35	15	38	1	1		18	4	8	1	2
Diplacodes haematodes	5	8	29	5	7		2	2	7	21	8	5	
Nannophlebia injibandi											1		
Nannophya fenshami		7											
Orthetrum caledonicum	11	9	28	12	17		1	4	9	11	7	7	
Orthetrum migratum	1	1	1	2	1						9		
Pantala flavescens		5	5		4			1	3	1			
Potamarcha congener										1			
Rhodothemis lieftincki											2		
Rhyothemis graphiptera		1								5	1		

	J	F	M	A	M	J	J	A	S	O	N	D	n.d.
Tholymis tillarga		1											
Tramea loewii		2							5	3	2		
Tramea stenoloba	1	2	2	3							1	3	
Zyxomma elgneri											1		

Temperate - Moderately dry Winter (hot Summer)

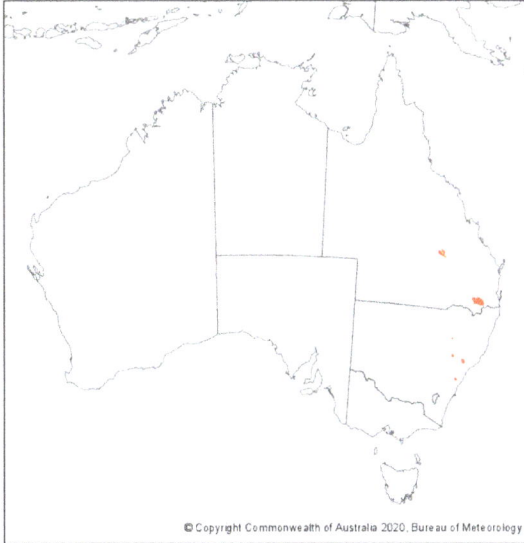

Temperate - moderately dry winter (hot summer)

Species   43   Specimens   202   Adults   Larvae

	J	F	M	A	M	J	J	A	S	O	N	D	n.d.
Family Synlestidae													
Episynlestes albicauda			3	1							1	1	
Synlestes selysi				1		2							
Synlestes weyersii			18								4		
Family Lestidae													
Austrolestes aridus					1								
Austrolestes leda	1			1					1				
Family Argiolestidae													
Austroargiolestes icteromelas	1	2	9	6					2			5	2
Family Lestoideidae													
Diphlebia coerulescens													1
Diphlebia lestoides												3	
Diphlebia nymphoides			3	1									
Family Coenagrionidae													
Austroagrion watsoni				14									

	J	F	M	A	M	J	J	A	S	O	N	D	n.d.
Ischnura aurora				2									
Ischnura heterosticta			1	2						2			
Pseudagrion aureofrons				1									
Xanthagrion erythroneurum			3										
Family Aeshnidae													
Adversaeschna brevistyla				2							1		
Austroaeschna muelleri			1								1		
Austroaeschna parvistigma		1											
Austroaeschna pinheyi			3	1									
Austroaeschna pulchra			5										
Austrogynacantha heterogena			1										
Dendroaeschna conspersa			3										
Hemianax papuensis	1	1							1				
Family Gomphidae													
Austrogomphus amphiclitus		1	4										
Austrogomphus cornutus		1											
Austrogomphus ochraceus										2			3
Hemigomphus gouldii			1								1	2	
Hemigomphus heteroclytus	1		6	2					1				
Family Synthemistidae													
Choristhemis flavoterminata			3									1	2
Eusynthemis aurolineata		1											
Eusynthemis deniseae			1								1	1	
Eusynthemis nigra			1										
Family Corduliidae													
Hemicordulia australiae			1										
Family Libellulidae													
Brachydiplax denticauda			1										
Crocothemis nigrifrons				1									
Diplacodes bipunctata				1	3				1				1
Diplacodes haematodes			3	4		1			3			1	
Nannodiplax rubra											1		
Nannophya dalei		1											
Orthetrum caledonicum		1	3	6					1				1
Orthetrum villosovittatum			4	4									
Zyxomma elgneri													1

Grassland - Hot (Summer drought)

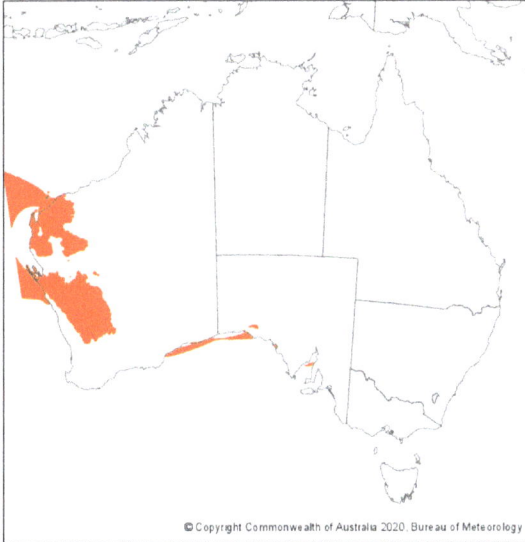

Grassland - hot (summer drought)

Species   41   Specimens   578   Adults   578   Larvae   0

	J	F	M	A	M	J	J	A	S	O	N	D	n.d.
Family Lestidae													
Austrolestes analis						3							
Austrolestes annulosus	1		4	3						3	2		1
Austrolestes aridus				1			3	1	1				
Austrolestes io						1							
Family Argiolestidae													
Miniargiolestes minimus												1	
Family Isostictidae													
Eurysticta coolawanyah		8								1	3	2	
Family Platycnemidae													
Nososticta pilbara	8	2					1			2	9	2	
Family Coenagrionidae													
Agriocnemis argentea	1	2		4			5			2	7	1	
Agriocnemis kunjina	1	3					2			1	13	4	2
Austroagrion pindrina	2						7				6	3	

307

	J	F	M	A	M	J	J	A	S	O	N	D	n.d.
Ischnura aurora	3		1	3	1		1			4	3	3	1
Ischnura heterosticta	2	3	1	4			2		4		7		
Pseudagrion aureofrons		3					5				3		1
Pseudagrion microcephalum	6	17		2			4			7	8	2	
Xanthagrion erythroneurum				2	1				2	4	4	1	
Family Aeshnidae													
Adversaeschna brevistyla	1				1								
Austrogynacantha heterogena			1										
Hemianax papuensis	6	1	1		2	5		1	1	4	3	1	3
Family Gomphidae													
Antipodogomphus hodgkini	54	7		1									1
Austroepigomphus gordoni	3	2								1	6	2	
Austrogomphus collaris													1
Austrogomphus mjobergi	2	6		3				2		8	2		1
Ictinogomphus dobsoni	9	7		1	1					1	2	5	
Family Corduliidae													
Hemicordulia australiae									1				
Hemicordulia intermedia			1	1									
Hemicordulia koomina	1						1				1	1	
Hemicordulia tau	4		1	1	3		1		1	8			1
Family Libellulidae													
Austrothemis nigrescens											1		
Crocothemis nigrifrons	4	3					2		4	1	3	1	
Diplacodes bipunctata	3	1	4	4	7		3		3		1		
Diplacodes haematodes	5	1		3	3	3	3		2	2	7		2
Macrodiplax cora					2				2	2	1	1	
Nannophlebia injibandi	4	6									2		2
Orthetrum caledonicum	9	5	1		5		2		1	2	5	1	1
Orthetrum migratum	4	2		2	1		2			1	4	1	
Pantala flavescens	1		2	1	11								
Rhodothemis lieftincki	3	2					1				2	3	
Rhyothemis graphiptera	3	2								1	4		
Tramea loewii			1	2									
Tramea stenoloba	1	2									1	1	
Zyxomma elgneri		2			1					2	2		

Temperate - Distinctly dry (and hot) Summer

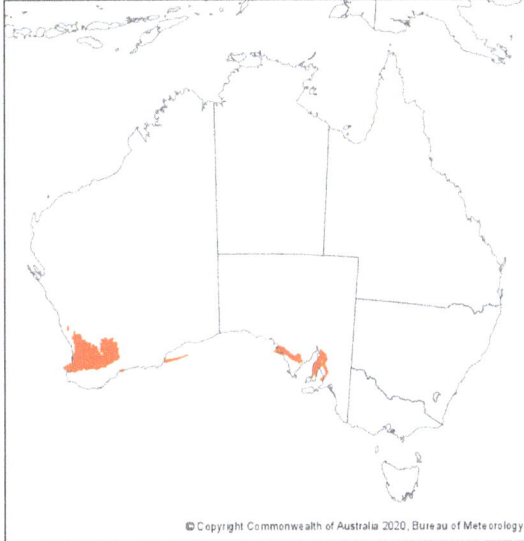

Temperate - distinctly dry (and hot) summer

Species 40   Specimens 925   Adults 925   Larvae 0

	J	F	M	A	M	J	J	A	S	O	N	D	n.d.
Family Lestidae													
Austrolestes aleison		10		1							4	5	
Austrolestes analis	2	2	2		1		1		1	14	11	8	2
Austrolestes annulosus	5	1	3	1	2				3	2	9	15	2
Austrolestes aridus								1		1			
Austrolestes io		1							2	1	6	7	
Family Argiolestidae													
Archiargiolestes parvulus										4	8	11	
Archiargiolestes pusillus	2								5	9	24	25	
Miniargiolestes minimus	11	2	1								4	19	1
Family Coenagrionidae													
Austroagrion cyane				4						22	3	7	2
Ischnura aurora	2		2							1	2	2	1
Ischnura heterosticta		1	4						1		5	5	4
Xanthagrion erythroneurum		1	4	1						3	2	7	2

309

	J	F	M	A	M	J	J	A	S	O	N	D	n.d.
Family Aeshnidae													
Adversaeschna brevistyla	8	1	1	1						9	8	9	3
Austroaeschna anacantha	9	4	1	1	1			2	2		4	13	
Austroaeschna parvistigma	1											1	
Austroaeschna unicornis	1	1	1										
Hemianax papuensis	2	2	4	3		1			1		2	6	3
Family Petaluridae													
Petalura hesperia	2								1		6		
Family Gomphidae													
Armagomphus armiger	2				1					13	9	6	
Austroepigomphus gordoni							1						1
Austrogomphus collaris	4	1								8	14	27	3
Austrogomphus guerini						1					1	1	
Zephyrogomphus lateralis	2										4	11	
Family Synthemistidae													
Archaeosynthemis leachii	5	3										11	
Archaeosynthemis occidentalis	4	1	1						1		9	15	
Archaeosynthemis spiniger	2	1									1		
Austrosynthemis·cyanitincta	1				1			1		4	8	20	1
Synthemis eustalacta	4											1	
Family Corduliidae													
Hemicordulia australiae	2								1		3	5	1
Hemicordulia tau	1	2	5	8	3			4	6	6	16	6	2
Procordulia affinis										5	8		1
Family Libellulidae													
Austrothemis nigrescens	5	3									13	7	
Diplacodes bipunctata	2	2						1	6		6	6	2
Diplacodes haematodes	5		2	2	1					8	5	5	4
Nannophya occidentalis									2	1	1	2	
Orthetrum caledonicum	6	1	2	4						2	23	12	3
Pantala flavescens			1										
Tramea stenoloba		1	1					1			3	1	1
Genera incertae sedis													
Hesperocordulia berthoudi	1	1								2	9	8	
Lathrocordulia metallica				1							1	4	

Grassland - Warm (persistently dry)

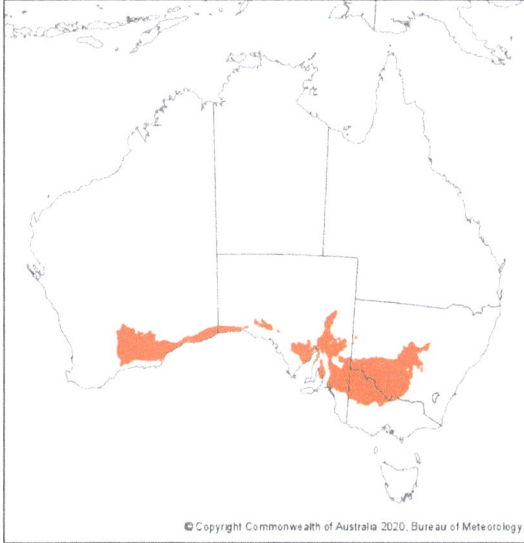

Grassland - warm (persistently dry)

Species   38   Specimens   1204   Adults   612   Larvae   592

	J	F	M	A	M	J	J	A	S	O	N	D	n.d.
Family Lestidae													
Austrolestes analis											3	1	1
Austrolestes annulosus	10	4	3	2		1		1		5	3	8	2
Austrolestes aridus	1			1				1	5		1		
Austrolestes leda	1		1		1			3			1	4	1
Austrolestes psyche			1							1			2
Family Lestoideidae													
Diphlebia nymphoides										1			
Family Isostictidae													
Rhadinosticta simplex		2											
Family Platycnemidae													
Nososticta solida		7									1	2	
Family Coenagrionidae													
Austroagrion cyane				1								3	
Austroagrion watsoni										1			

	J	F	M	A	M	J	J	A	S	O	N	D	n.d.
Ischnura aurora	5	13	2	4		1	1	2	2	9	3	5	1
Ischnura heterosticta	11	13	2	5	1				1	3	2	3	1
Pseudagrion aureofrons		8								1		2	
Xanthagrion erythroneurum	10	32	8	6	3	1				12	5	27	3
Family Aeshnidae													
Adversaeschna brevistyla		2	1						1	1	2	1	
Austroaeschna multipunctata		1										1	
Austroaeschna parvistigma		1		3							1	3	
Austroaeschna unicornis		1											
Hemianax papuensis	2	3	9		1			2	3	2	4	5	3
Telephlebia brevicauda		1											
Family Gomphidae													
Austroepigomphus praeruptus													1
Austrogomphus angelorum											2	2	
Austrogomphus australis	5	3	4								6	22	1
Family Synthemistidae													
Eusynthemis brevistyla		1											
Synthemis eustalacta		1											
Family Corduliidae													
Hemicordulia australiae											1		
Hemicordulia tau	2	5	8	6	2	1			5	10	10	5	4
Family Libellulidae													
Austrothemis nigrescens	1												
Crocothemis nigrifrons	1												
Diplacodes bipunctata	11	16	7	5	4	1	2		5	12	7	6	1
Diplacodes haematodes		6			1	1			1	2		3	1
Orthetrum caledonicum	15	18	4	1	1		1		1	10	3	2	3
Pantala flavescens			2										
Tramea loewii			1										

Subtropical - Distinctly dry summer

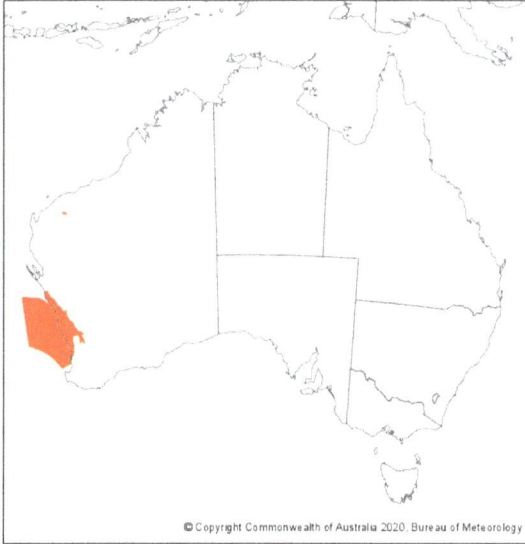

Subtropical - distinctly dry summer

Species 35   Specimens 1086   Adults 1086   Larvae 0

	J	F	M	A	M	J	J	A	S	O	N	D	n.d.
Family Lestidae													
Austrolestes aleison								2	1	2		2	
Austrolestes analis	7	4	2	3		1			6	30	4	10	1
Austrolestes annulosus	7	2		2	2	1	1		8	8	12	7	3
Austrolestes aridus								1	2	3			
Austrolestes io	3		1	2		3	1	3	3	11		2	
Family Argiolestidae													
Archiargiolestes parvulus										2	22	1	
Archiargiolestes pusillus	1								3	20	24	7	
Miniargiolestes minimus	4	1		1							12	5	
Family Coenagrionidae													
Austroagrion cyane	1			1					6	7	4	4	4
Ischnura aurora	4	1	1			1			1	9	1	6	
Ischnura heterosticta									2				
Xanthagrion erythroneurum	7	2	2	2					2	2	1	15	4

313

	J	F	M	A	M	J	J	A	S	O	N	D	n.d.
Family Aeshnidae													
Adversaeschna brevistyla	10	5	3	1	1	2		1	5	8	9	4	1
Austroaeschna anacantha	7	1	2		1						4	4	
Austrogynacantha heterogena	1												
Hemianax papuensis	10	8	4	10	2	11	1	1	1	4	5	14	2
Family Petaluridae													
Petalura hesperia	4										1	30	
Family Gomphidae													
Austrogomphus collaris	6	1			1					25	19	12	9
Zephyrogomphus lateralis	4										2	5	
Family Synthemistidae													
Archaeosynthemis leachii	3										3	4	
Archaeosynthemis occidentalis	2									1	10	11	
Austrosynthemis·cyanitincta											5	2	
Family Corduliidae													
Hemicordulia australiae	2	1	6	1	8				1	8	7	4	
Hemicordulia tau	7	4	2	8	12	12		2	13	18	9	11	1
Procordulia affinis	1							7	4	2	1		
Family Libellulidae													
Austrothemis nigrescens	9		2	1				1		8	12	22	1
Diplacodes bipunctata	10	10	4	2	2			1	1	15	11	21	
Diplacodes haematodes	3	3	2					2	3	3	2		7
Nannophya occidentalis			1							1	1		
Orthetrum caledonicum	6	7	7	2					3	4	11	32	8
Pantala flavescens			5										
Tramea loewii			1										
Tramea stenoloba	1	1	5							1	3	4	
Genera incertae sedis													
Hesperocordulia berthoudi	2		1							1	6		
Lathrocordulia metallica												1	

Desert - Hot (persistently dry)

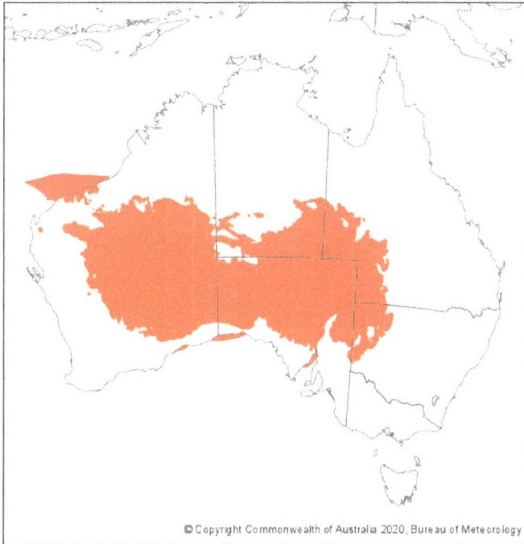

Desert - hot (persistently dry)

Species   28   Specimens   448   Adults   403   Larvae   45

	J	F	M	A	M	J	J	A	S	O	N	D	n.d.
Family Lestidae													
Austrolestes annulosus	1			1		1			1	1	2		
Austrolestes aridus		1	1				2	2	4	4	6		
Austrolestes leda							1		1				
Family Isostictidae													
Eurysticta coolawanyah											1		
Family Coenagrionidae													
Agriocnemis argentea								1					
Agriocnemis kunjina											1		
Argiocnemis rubescens										1			
Austroagrion cyane			1										
Austroagrion pindrina											1		
Austrocnemis maccullochi										2			
Ischnura aurora	3		1		1			1		14	12		
Ischnura heterosticta		3			3				7	4	3	3	1

	J	F	M	A	M	J	J	A	S	O	N	D	n.d.
Pseudagrion aureofrons										1		2	
Xanthagrion erythroneurum	3	2	2			1		1	4	11	12		2
Family Aeshnidae													
Adversaeschna brevistyla						1			1				
Austrogynacantha heterogena					1								
Hemianax papuensis		3	7	3	7		3	1	5	1	9	1	4
Family Gomphidae													
Austroepigomphus gordoni									2				
Austrogomphus australis										3	2	2	1
Family Corduliidae													
Hemicordulia flava		1											
Hemicordulia tau	4	3	7	8	7		2		5	2	4	5	6
Family Libellulidae													
Diplacodes bipunctata		5	7	1	15	5		8	1	2	5	4	2
Diplacodes haematodes		10	2		3	1	1		9	2	5		1
Macrodiplax cora					1								
Orthetrum caledonicum		7	4	5	3	1	1	1	6	3	20	3	1
Orthetrum sabina					1								
Pantala flavescens		3	6		1								
Tramea loewii		1						1					

Temperate - No dry season (cool Summer)

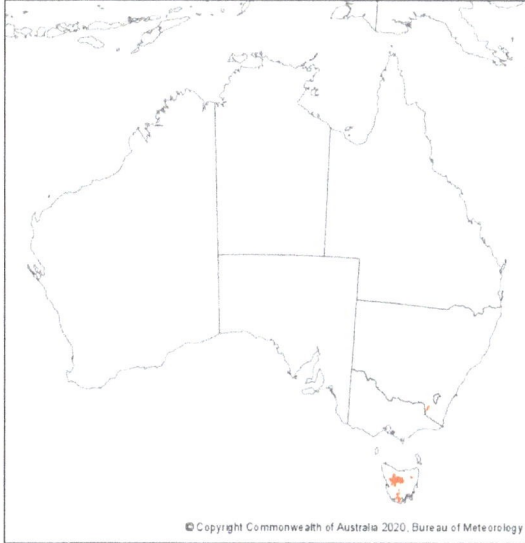

Temperate - no dry season (cool summer)

Species 23   Specimens 206   Adults 206   Larvae 0

	J	F	M	A	M	J	J	A	S	O	N	D	n.d.
Family Lestidae													
Austrolestes analis		2											
Austrolestes annulosus	2	2	1									1	
Austrolestes cingulatus	4	5										9	
Austrolestes psyche	7	11										6	
Family Argiolestidae													
Austroargiolestes calcaris	1												
Griseargiolestes intermedius	2												
Family Coenagrionidae													
Ausrocoenagrion lyelli												1	
Ischnura aurora		2											
Ischnura heterosticta	1												
Family Austropetaliidae													
Archipetalia auriculata	4											5	2

	J	F	M	A	M	J	J	A	S	O	N	D	n.d.
Family Aeshnidae													
Austroaeschna flavomaculata	4	1											
Austroaeschna hardyi	17	11											
Austroaeschna inermis	4												
Austroaeschna parvistigma	2	17	1								1		
Austroaeschna tasmanica	1	4	1							1			
Telephlebia brevicauda	2												
Family Gomphidae													
Austrogomphus guerini	1									1			
Family Synthemistidae													
Eusynthemis guttata	3												
Syn. gomphomacromioides	7	5											1
Synthemis eustalacta	2												
Synthemis tasmanica	12	27	1								2	2	
Family Corduliidae													
Hemicordulia tau		2	1								2		
Procordulia jacksoniensis	1	1											

www.ingramcontent.com/pod-product-compliance
Lightning Source LLC
Chambersburg PA
CBHW052108030426
42335CB00025B/2894